THE BEGINNER'S GUIDE TO

Astronomy

THE BEGINNER'S GUIDE TO

Astronomy

GALLERY BOOKS
An imprint of W. H. Smith Publishers Inc.
112 Madison Avenue
New York, New York 10016

To Susan and Matthew

Executive Managers	Kelly Flynn
	Susan Egerton-Jones
Editor	Fred Gill
Art Editor	Ruth Levy
Editorial Assistant	Stephen Bowden
Production	Peter Philips

Edited and designed by the
Artists House Division of
Mitchell Beazley International Ltd
Artists House
14–15 Manette Street
London W1V 5LB

"An Artists House Book"

ISBN 0-8317-0745-3

Typeset by Hourds Typographica, Stafford.
Reproduction by La Cromolito s.n.c., Milan.
Printed in Portugal by Printer Portuguesa, Lisbon

Foreword

Following in the wake of Halley's Comet, guides to astronomy have poured into the bookshops at the speed of a stream of shooting stars. The result is that you can travel the Universe without ever having to leave the comfort of your armchair. Most of these guides – guides to the planets, the stars, comets and meteors, the Universe itself – are studded with sensational photographs, captured by probes far out in space and enhanced with the latest in computer wizardry. They are packed with mind-boggling descriptions of exotic objects unknown to astronomers only a generation ago. Books like these stir the imagination: they take the mind away from the sometimes mundane process of living on Earth to the humbling vastness of the furthest reaches of the Universe.

But the coming of the comet has produced more than just a book explosion. It has given rise to a whole new band of enthusiastic sky-watchers: people who, not content with armchair explorations, want to go out and get to know the starry sky for themselves. But their problem is: where do you start? There is so much to see – and to learn. For instance, on a clear, dark night, there are 3,000 stars on show; how can anyone tell one from another? Then there are planets: interlopers that move from night to night against the starry background. Meteors – "shooting stars" – can catch a sky-watcher by surprise at almost any time. The Moon comes and goes during the month as it follows its cycle of phases, and – just occasionally – it may disappear into eclipse. And just when the beginning astronomer thinks that he or she has grown familiar with the stars of one season, an entirely new parade of constellations slowly rolls in to replace them as the months slip by. Whoever wrote that the heavens are unchanging can never have looked up!

What the beginning observer needs is an experienced guide: and Brian Jones' *The Beginner's Guide to Astronomy* is the ideal companion. First and foremost, it is a *practical* book. Here is all the basic information you need to get started as an amateur astronomer. There are simple, yet detailed starcharts, pointing out the "tourist spots" in all the main constellations. There's advice on *how* to observe, from someone who has weathered many freezing nights gazing heavenward from the back garden. When it comes to the pros and cons of choosing a telescope or binoculars, Brian Jones offers sound, experienced advice. And for those with ambitions to build an observatory, look no further: the details are all here, right down to the instructions for making the concrete floor or base.

Stargazing as a hobby is fun, rewarding and (reasonably!) cheap. But it can also be genuinely useful. In these high-tech times, it's ironically true that professional astronomers – whose observing targets have become highly specific – are relying increasingly on amateurs to fulfil the role of monitoring the sky for unexpected changes. Hundreds of amateur astronomers have made important discoveries with equipment no more sophisticated than a pair of binoculars. With *The Beginner's Guide to Astronomy* as your companion, you will soon feel at home in the heavens – and perhaps, one day, join the ranks of the discoverers themselves.

HEATHER COUPER
Vice-President, British Astronomical Association, and President, Junior Astronomical Society

Contents

KEY TO SYMBOLS

- Variable Stars
- Nebulae
- Star Clusters (Open)
- Star Clusters (Globular)
- Planetary Nebulae
- Irregular Galaxies
- Other Galaxies

The Unfolding Stars

Since before recorded history man has gazed at the heavens with wonder, watching the seasonal changes in the star patterns and following the paths of the planets as they travelled across the sky. To our distant ancestors the heavens took on the appearance of a gigantic hollow sphere, suspended from which were countless tiny lamps. We have progressed greatly from then and the universe is now known to be overflowing with strange and mysterious objects, a seemingly boundless cosmos playing host to countless galaxies each containing billions of stars.

The night sky is certainly a fascinating place and can be a source of enjoyment to anyone who is prepared to simply go out and have a look. The amateur astronomer can also play an important role and many notable discoveries have been made by members of this rapidly growing band of enthusiasts. There is nothing quite as stimulating as gazing up at a dark and moonless sky. We are children of the stars, our planet and all things on it forming through the collecting together of elements produced during a supernova explosion aeons ago. Perhaps, therefore, part of the wonder of looking up into the night sky is the thought that we are looking back at our roots.

The starry sky looks much the same to us as it did to the astronomers who lived thousands of years ago and we can pick out the patterns they devised as they mapped out the firmament. Perseus and Andromeda still command our autumn sky, while King Cepheus and Queen Cassiopeia hold hands on the banks of the Milky Way. Each of these constellations is made up of a number of different stars, some of which appear bright while others hover on the point of naked-eye visibility. One of the brightest stars is Arcturus in the constellation of Boötes, the Herdsman. Arcturus is fairly close to us, its light taking around 35 years to reach our planet. On the other hand Rigel, marking one of the feet of Orion the Hunter, shines from a distance of 900 light years. Each and every star that we see in our night sky is just one of over 100,000 million that make up the giant spiral-shaped system of stars we know as the Milky Way Galaxy. This in turn is just one of around 100 million other galaxies scattered around the universe. If all this seems rather grand, ponder for a moment an object known as PKS 2000-330. This is a quasar, an extremely distant object which emits a fantastic amount of energy and which lies at the very edge of the observable universe. Its light has been travelling towards us for some 15,000 million years, setting off almost three times the age of the Earth ago!

However, take heart! These vast distances and tremendous time scales need not overawe you. With only moderate optical aid, or even the naked eye, many of the colourful and exciting objects that exist throughout space await your attention. A dark night, the *Beginner's Guide to Astronomy* and a little patience are all you need to join the happy band of astronomers who constantly derive great pleasure and enjoyment from the night sky. Astronomy is certainly the most emotionally rewarding hobby of all, and your first glance may spark off a romance that could last a lifetime. Happy stargazing!

Many people have helped in the compilation of this book. I would particularly like to thank Robin Scagell, Paul Money, John Fletcher and Derek Hufton for their invaluable assistance with photographs and thanks also to George Haig for the Scotch Mount reference. Valuable information on variable stars was received from Colin Henshaw, John Isles, Doug Saw and Dr. Janet A. Mattei, while the generally useful inspiration provided by Steve Bell and Neville Kidger did not go unnoticed! My thanks also to Ruth Levy, Susan Egerton-Jones and Kelly Flynn for their help and encouragement.

BRIAN JONES

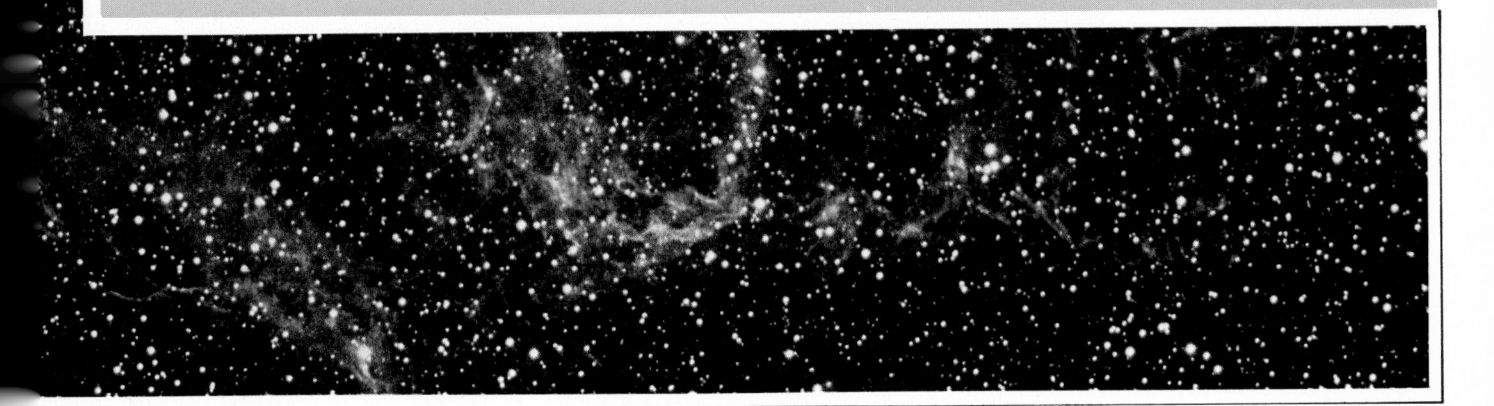

Telescopes

There are two basic types of astronomical telescope – refractors using a lens to collect the light and bring it to a focal point, reflectors having a specially shaped mirror for this purpose. The earliest telescopes were of the refracting type, the original refractors being based on the crude instruments of Galileo and his contemporaries.

Although these early telescopes revolutionized optical astronomy, the images they produced were sometimes greatly distorted. The trouble was that the lenses (objectives) had a single component and produced chromatic aberration whereby the different wavelengths of light, ranging from red through to violet, were brought to focus at different points. This tended to create a great deal of false colour in the image. In 1758 the English optician John Dollond, together with his son Peter, perfected the achromatic objective, which greatly helped to reduce the problems of false colour.

The first reflecting telescope was developed and built by Isaac Newton in 1668, although it must be said that the idea of using a mirror instead of a lens to gather light was not entirely new. Newton's first telescope was very small, although the design was simple and straightforward, so much so that telescopes based on the Newtonian principle are by far the most popular among amateurs today.

When it comes to choosing a telescope the advantages and disadvantages of each system must be weighed up, and one of the prime considerations is that of cost. Refractors are generally more expensive size for size than reflectors because there is far more work involved in figuring the many optical surfaces in an achromatic lens than in the single reflecting surface of a mirror. The aperture should receive equal consideration. This is the diameter of the object glass in the case of a refractor, or of the primary mirror in the case of a reflecting telescope. The larger the aperture of the telescope, the more light it can gather, and consequently the fainter the objects that can be observed.

The magnification of a telescope depends upon the eyepiece. Here, the value is obtained by dividing the focal length of the eyepiece into that of the lens or mirror. The focal length is equal to the distance between the objective and the point at which the image is produced, and is expressed as a ratio of the objective diameter. For example, a 6-inch f/6 reflector has a focal length of $6 \times 6 = 36$ inches and when used with an eyepiece with a focal length of $\frac{1}{2}$ inch would produce a magnification of $36/\frac{1}{2}$ or $72\times$. However, when high magnifications are being used, the field of view, or area being observed, is relatively small. The smaller this area the less light it gives out. So, large-aperture telescopes are needed in order to collect this light and make observing with high powers practical.

Bearing in mind the aperture, telescopes of long focal lengths allow high magnifications to be obtained. Generally speaking, refractors have long focal ratios, making them more suited for detailed work such as planetary observing. Reflectors, with their relatively short focal lengths, are more suitable for deep-sky observing, their wide fields of view being ideal for the study of nebulae, star clusters and galaxies.

When used in conjunction with astronomical eyepieces, virtually all types of telescope give an inverted image, which, to the astronomer, is no real disadvantage. In order to produce an upright image additional lenses would be needed in the eyepiece, although each of these lenses would absorb some light and would therefore produce a slightly fainter image. Erect-

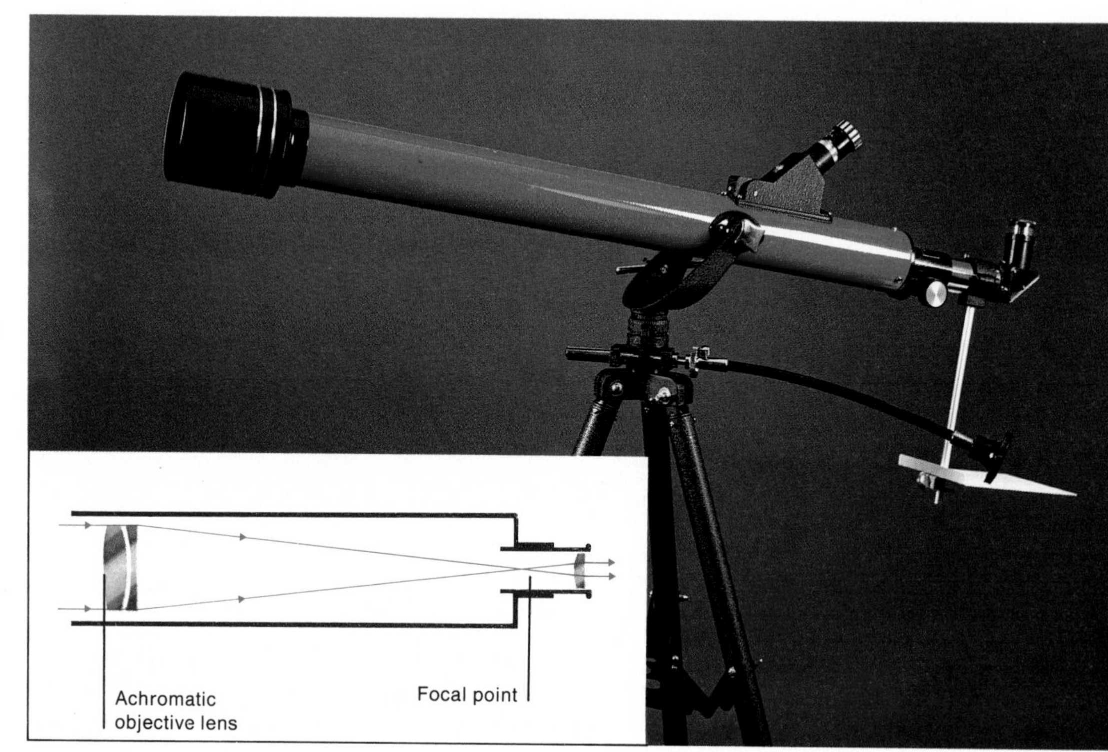

Right: *a 60 mm-aperture refracting telescope on an alt-azimuth mount. Note the solar projection screen and the star diagonal situated at the eyepiece end of the telescope. The star diagonal can help greatly when viewing at awkward angles.*
Lower right: *in a refracting telescope light enters the system through a lens after which it is bent, or refracted, to a focal point where the image produced is magnified by an eyepiece. Achromatic objectives are made up of a number of different components, each constructed from a different type of glass. These collectively minimize the effects of chromatic aberration, although the problem cannot be completely eliminated.*

TELESCOPE MOUNTS

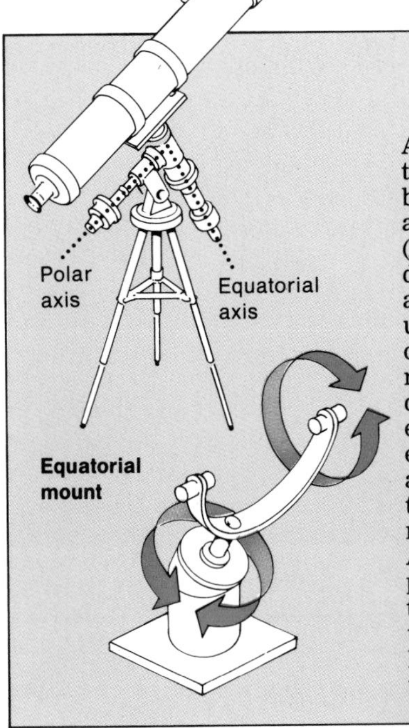

Alt-azimuth mounts enable the telescope to be moved both vertically (*ALT*itude) and horizontally (*AZIMUTH*). The main disadvantage is that both axes of the mounting must be used in order to keep the star or other object within the narrow field of view. These difficulties can be largely eliminated by the use of an equatorial mounting which allows the observer to keep the star in the field of view by movement in one axis only. An equatorial mounting has polar and declination axes, the polar axis being fixed so that it lies parallel to the Earth's axis and pointing in the direction of the celestial pole. Once this initial alignment is complete the telescope is brought to bear on the chosen star by moving the telescope around both axes, after which the declination axis is clamped down. The telescope can now be made to follow the star as it crosses the sky by turning the telescope around the polar axis only.

Although equatorial mounts make prolonged observation easier, tracking over a long period of time can be tiring. This problem can be overcome by attaching an automatic drive to the mount, which compensates for the daily rotation of the Earth.

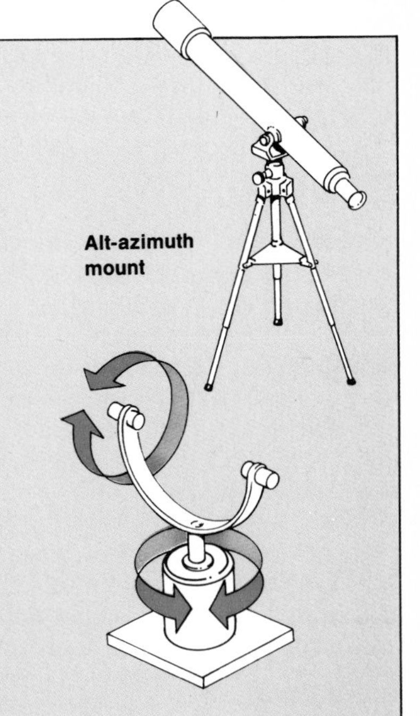

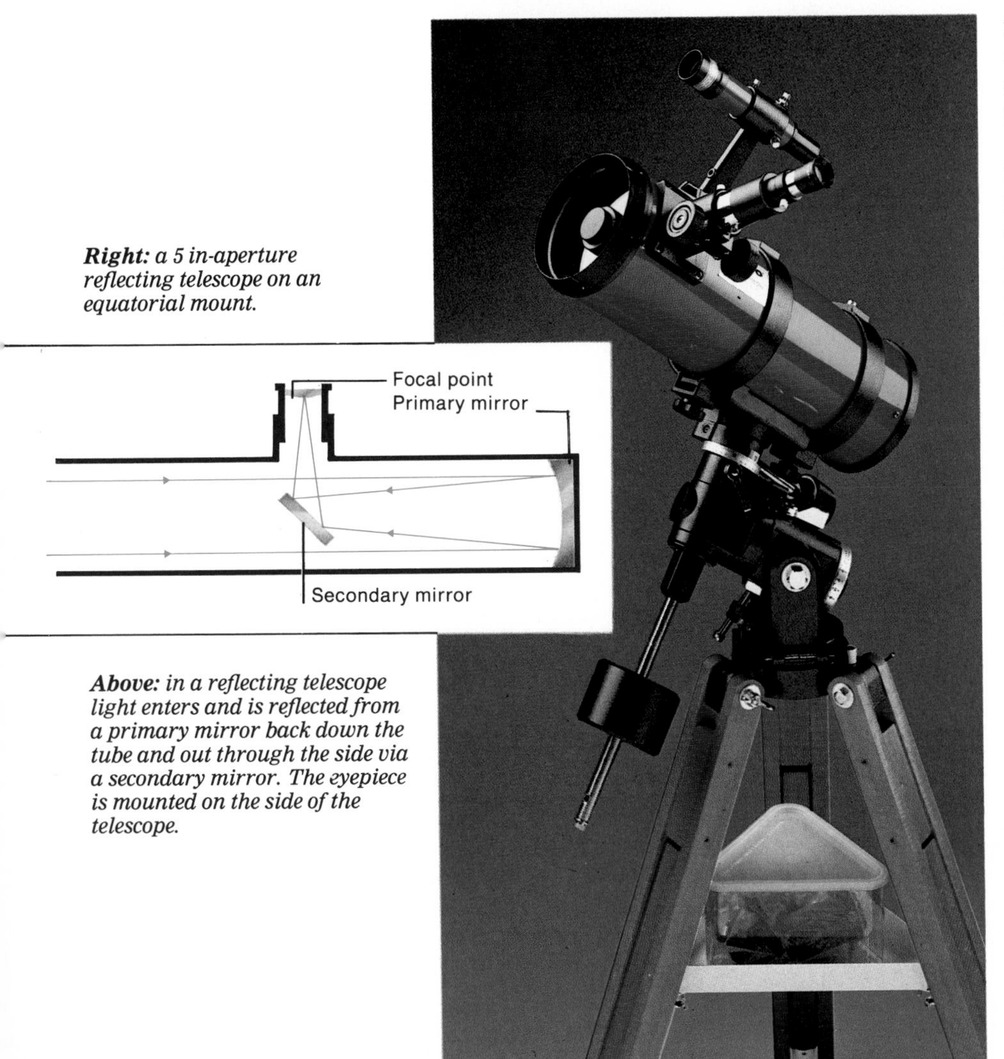

Right: *a 5 in-aperture reflecting telescope on an equatorial mount.*

Above: *in a reflecting telescope light enters and is reflected from a primary mirror back down the tube and out through the side via a secondary mirror. The eyepiece is mounted on the side of the telescope.*

ing prisms are available that make the object in view the right way up, although they are seldom used except for terrestrial viewing.

Large telescopes should be permanently sited. Reflectors are generally more compact, and are easier to carry around. However, if the aperture of a refracting telescope is 4 in (10 cm) or more (8 in/20 cm) or more for a reflector) it should really be housed in some form of observatory.

Whatever the type and size of telescope it should be regularly cleaned and all moving parts lubricated and periodically checked for wear and tear. With reflectors the mirror may require occasional realignment and the reflective coating on the surface of the mirror will need to be renewed from time to time. This coating takes the form of a very thin layer of aluminium and replacing it is really a job for the specialist. When a reflecting telescope is not in use both the primary and secondary mirrors should be covered to prevent dust and grease settling on them. If this simple precaution is carried out the useful life of the mirror coatings will be greatly extended.

Binoculars

Binoculars are inexpensive, easy to use and are without doubt the best type of instrument for the first-time observer. Lunar features, planets and their satellites, comets and a whole host of deep sky objects, including double stars, variable stars, star clusters, nebulae and galaxies, are all within the reach of binoculars. They are fairly straightforward in construction and basically take the form of two small refracting telescopes held firmly together and aligned with each other. Most binoculars contain prisms, which fold up the light from the object under observation, thereby producing a more compact instrument. The prisms also bring the two images closer together so that they emerge at roughly the same distance apart as the eyes of the observer. Final adjustment is made by swinging the two barrels around a central hinge or pivot to vary the distance between the eyepieces.

Above: *binoculars are the ideal instrument for the beginner.* ***Left:*** *the 8 × 56 binoculars shown here will produce some excellent images of the night sky, their large objectives giving the light grasp essential for astronomical work.*

Choosing Binoculars

Many different makes of binocular are available and choosing a good pair at a price to suit your pocket can seem a daunting task. Binoculars are designated as A × B, where A is the magnification and B is the aperture. It must be borne in mind that the higher the magnification, the harder it will be to hold the instrument steady while observing. Opinions vary as to the maximum power for hand-held binoculars and the limit seems to be somewhere between 7× and 10×. The aperture is the most important consideration and should be as large as possible, the governing factor being that very large lenses will mean that the binoculars are heavy and therefore difficult to hold steady for prolonged periods. Generally speaking, 7 × 50 binoculars seem to be the most preferable for astronomical purposes, giving a good magnification, plenty of light-gathering power and a fairly wide field of view.

Whether new or second-hand binoculars are purchased, there are a few simple tests that can be carried out which will help ensure that money isn't wasted. Pick up the binoculars and examine the general standard of workmanship. Then fully extend the eyepieces and make sure that they are rigid and don't tilt. Both the eyepiece focusing mechanism and the central pivot should operate smoothly but with steady resistance.

In most binoculars the lenses have anti-reflection coatings which help to increase both contrast and light transmission, and in good-quality instruments all the optical surfaces should be coated. To determine whether the optics are coated hold the binoculars in front of you so that they reflect a bright light source shining over your shoulder. There will be two reflections from the large objective, one from the front of the lens and one from the back. Coloured reflection(s) show that the optical surfaces are coated. Now move the binoculars around until you see a similar reflection from the prisms inside the barrel. In each case, a white reflection will indicate a surface without a coating. The examination can be repeated for the eyepiece end by turning the binoculars round. Instruments with no coating whatsoever are not to be recommended.

Now focus the binoculars on a straight line, such as a telephone wire or length of guttering. At the edge of the field of view the line will be seen to bend. Although a small amount of distortion is to be expected it should not be excessive. Another test of optical quality can be carried out by looking at a star or, if that isn't possible, at something similar, such as the reflection of sunlight from a window or other shiny surface. Centre the object in the field of view, bring it to focus and then move it slowly towards the edge of the field. By the time it reaches the edge it will probably have become out of focus, although it should remain as a distinct point until at least half-way out from the centre. However, if the binoculars have what is termed a flat field the star should remain in focus regardless of its position. Finally, check that the binoculars don't produce a double image, a fault that will arise if the barrels are not set parallel to each other. This can be a little difficult to detect because your eyes will tend to automatically compensate for any misalignment. The best method is to look through the binoculars, closing and reopening your eyes, carefully checking for a double image before your eyes adjust. Any instrument that displays a fault of this nature should be discarded.

A set of lens caps and a carrying strap are essential pieces of equipment. The caps will keep dust and grease off the objectives and eyepieces when the binoculars are not in use. The carrying strap must always be worn when the binoculars are being used as a safety measure in case they are dropped. Another useful addition would be a tripod, together with a binocular adaptor, which will enable the instrument to be mounted and held steady during use. This will greatly enhance their performance.

Don't be discouraged if you can't immediately find a suitable pair of binoculars, or if the ones you do purchase are not perfect in every detail. Unless you pay a lot of money, few are. However, if time is taken in making your choice, this will be amply rewarded with many years of enjoyable observation.

BINOCULAR TYPES

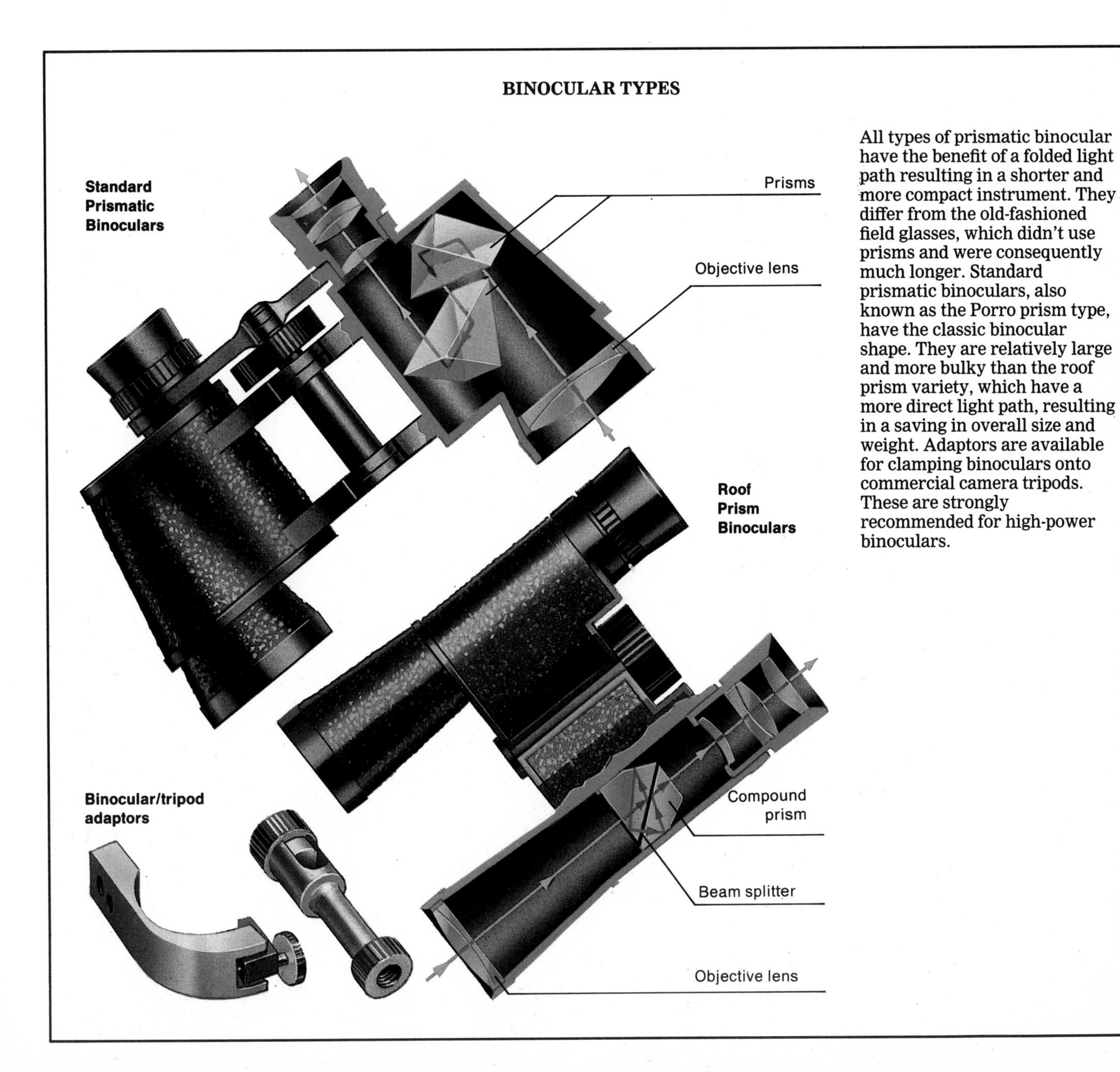

All types of prismatic binocular have the benefit of a folded light path resulting in a shorter and more compact instrument. They differ from the old-fashioned field glasses, which didn't use prisms and were consequently much longer. Standard prismatic binoculars, also known as the Porro prism type, have the classic binocular shape. They are relatively large and more bulky than the roof prism variety, which have a more direct light path, resulting in a saving in overall size and weight. Adaptors are available for clamping binoculars onto commercial camera tripods. These are strongly recommended for high-power binoculars.

The Solar System

Many different theories have been advanced to explain how the Solar System came into being. One of the first was the Nebular Hypothesis, proposed by the French astronomer Pierre Simon de Laplace in 1796. Laplace suggested that the Sun and planets formed from a rotating cloud of gas and dust from which a series of rings were thrown off. These rings eventually condensed to form the planets, while the central region of the cloud contracted to form the Sun. Closer examination, however, proved that the cloud could not have been rotating fast enough for rings of material to have been thrown off as suggested.

At around the turn of the century it was the opinion of some astronomers that the planets had formed from either a cloud or filament of material that was dragged from the Sun by a passing star. According to one theory the giant planets formed from the central regions of the filament, the smaller planets taking shape at either end.

Whatever the processes were that gave birth to the Solar System, astronomers are agreed that formation took place around 4,500 million years ago. This has been verified through the study of meteorites, in particular a dark, rocky type of meteorite, samples of which have fallen to Earth from interplanetary space. These meteorites represent the oldest type of rock known, and are thought to be essentially unchanged since their formation. Their age has been put at around 4,500 million years, a good indication that the Solar System came into being at this time.

Astronomers also agree that the Sun and planets evolved from an interstellar cloud of gas and dust. These clouds are known as nebulae, and many examples can be seen today, scattered throughout our own and other galaxies. Yet, although we know the type of environment in which the Solar System was born, the finer details of how the formation took place still cause disagreement. Observation of the general characteristics of the Solar System offers certain clues as to the processes of formation. For instance, all the planets orbit the Sun in more or less the same plane and are also seen to be split into two groups; the four inner planets being small, rocky worlds, while the next four are gaseous in composition and have much larger diameters. The one anomaly seems to be Pluto, which is, as far as we can tell, a small rocky planet travelling round the Sun in a highly eccentric orbit that is greatly inclined to the orbits of the other planets.

It is generally believed that the Sun and planets did indeed evolve from a spinning cloud of gas and dust. As it rotated, the cloud gradually became flattened, with the majority of its mass being concentrated towards the centre. This concentration of mass eventually contracted to form the Sun.

The planets were formed in the surrounding disc of material. The lighter, faster-moving elements in the disc were driven away by the Sun's heat from the inner Solar System, in which only dense, slow-moving material could survive to form the terrestrial planets. The lighter elements, which took the form of ice particles, eventually collected together to form the four gaseous planets. Any remaining material scattered throughout the space between the planets was quickly removed by the constant stream of energized particles emitted by the Sun. This stream of particles, known as the solar wind, can still be detected today. Modern research has shown us that Laplace was not too far from the truth when he advanced his Nebular Hypothesis almost two centuries ago.

The Asteroids

In 1766 Johann Titius publicized an interesting numerical relationship linking the distances of the planets from the Sun. Taking the numbers 0, 3, 6, 12, 24, 48 and 96, each one of which, apart from 3, has a value twice that of the previous number, he added 4 to each to give the sequence 4, 7, 10, 16, 28, 52 and 100. Taking the value of the Earth's distance from the Sun as 10 it was found that the distances of the other known planets fell into place remarkably well. There was, however, no known planet to correspond with the value of 28 and Titius suggested that there must be an undiscovered body orbiting the Sun between Mars and Jupiter. A look at any scale diagram of the Solar System will

Pluto

Neptune

Uranus

Of all the planets, Pluto has the longest (248 years) and most eccentric orbit. It also has an orbital inclination of 17°, the largest in the Solar System. For part of its journey around the Sun, near perihelion, Pluto crosses the orbital path of Neptune and for a while gives up the role of "outermost planet". The last such crossing took place in 1979, and until 1999 Neptune will actually be the furthest planet from the Sun.

indeed show a marked gap between the orbits of these two planets. At first, astronomers were rather sceptical, even when Johann Bode revived the idea in 1772. He publicized it so much that it came to be known, perhaps unfairly, as Bode's Law. However, when Uranus was discovered in 1781 it was found to tie in with the law, which gave a distance of 196 as compared with the actual value of 191.8.

At last astronomers began to take notice, so much so in fact that in 1800 Johann Schroter called together a group of observers with the intention of searching for the missing planet. This group christened themselves the "Celestial Police" and included among their ranks the Hungarian Franz Xaver von Zach and the Germans Heinrich Olbers and Karl Harding. However, the first asteroid was discovered quite independently by Giuseppe Piazzi from Palermo in Sicily, on 1 January, 1801. The discovery was made accidentally while Piazzi was compiling a star catalogue, although its distance from the Sun was found to tie in very well with Bode's Law with a value of 27.7. Ceres, as the newly discovered planet came to be known, turned out to be rather small and the Celestial Police continued their search, which culminated in the discovery of Pallas by Olbers in 1802. Karl Harding discovered Juno in 1804, and 1807 saw the discovery of Vesta, again by Olbers. After Vesta no more discoveries were forthcoming and the Celestial Police disbanded in 1815.

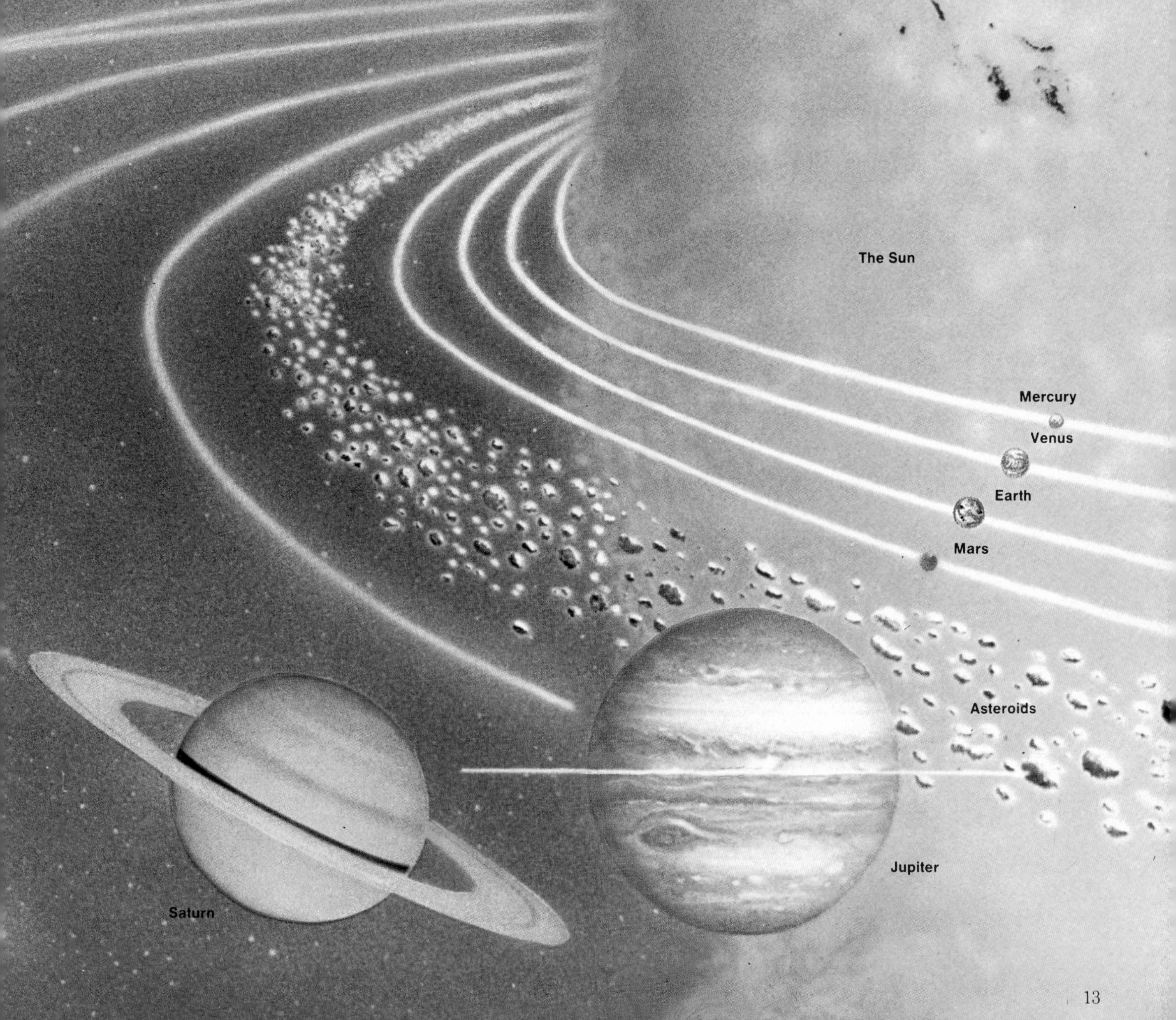

Interest in the asteroids, or minor planets, suffered a decline, although in 1830 the Prussian astronomer Karl Hencke began a further search and his efforts were rewarded by his discovery of Astraea in 1845. Hencke also brought to light the sixth minor planet, Hebe, in 1847, and since then not a year has gone by without further discoveries being made. For most of the nineteenth century, asteroids were detected by the visual method, which entailed many hours of work at the eyepiece end of a telescope checking and rechecking star fields, looking for tiny points of light that would betray their true identities through their movements against the background of stars. However, in 1891 a major turning point in minor planet discovery was reached.

The man responsible for a huge increase in the number of known asteroids was Maximilian Franz Joseph Cornelius Wolf. Better known as Max Wolf he regarded astrophotography as an important part of his work and in 1891 he made the first photographic discovery of a minor planet. The method was simple in that a camera was made to follow the stars. Any object with its own motion through the sky would show up as a streak on the photograph, unlike the stars, which would show as points of light. In all, Wolf discovered 232 asteroids, to which permanent numbers have been given.

As the twentieth century progressed, more and more asteroids were discovered and their numbers grew so rapidly that by 1923 over a thousand had been assigned permanent numbers. Their images appeared on photographic plates that had often been taken as part of completely unrelated research and many astronomers came to regard them as a positive nuisance. However, the minor planets are turning out to be very interesting, and the study of these bodies will, we believe, tell us much about the early history of the Solar System, and may unlock long-hidden secrets of our celestial neighbourhood.

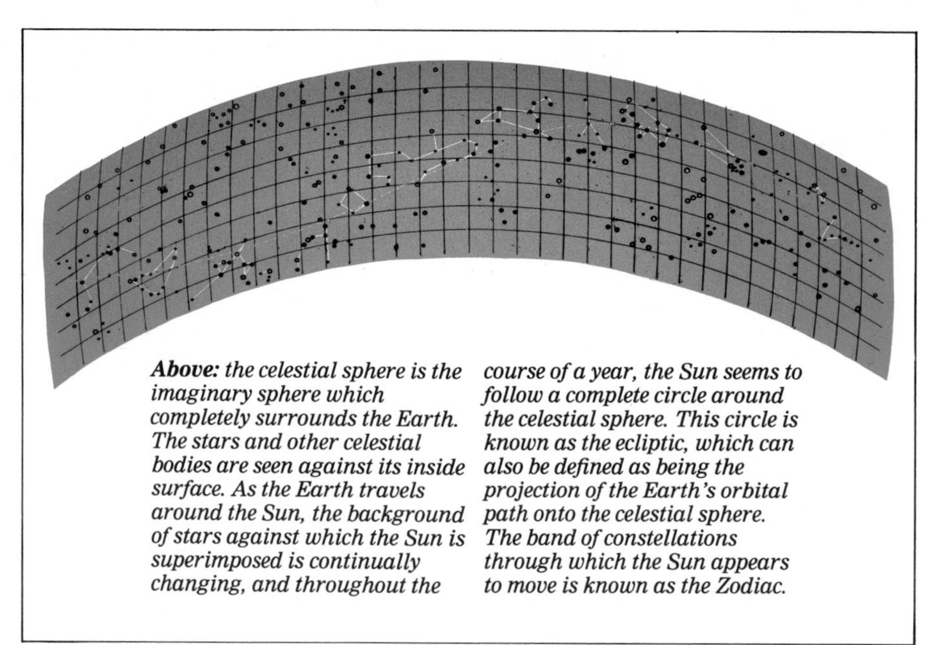

Above: *the celestial sphere is the imaginary sphere which completely surrounds the Earth. The stars and other celestial bodies are seen against its inside surface. As the Earth travels around the Sun, the background of stars against which the Sun is superimposed is continually changing, and throughout the course of a year, the Sun seems to follow a complete circle around the celestial sphere. This circle is known as the ecliptic, which can also be defined as being the projection of the Earth's orbital path onto the celestial sphere. The band of constellations through which the Sun appears to move is known as the Zodiac.*

Pluto

Discovered by the American astronomer Clyde Tombaugh in 1930 and named after the Guardian of the Underworld, Pluto is the outermost planetary member of the Solar System.

Pluto takes 248 years to orbit the Sun at a mean distance of 3,666 million miles (5,900 million km). Its diameter is extremely difficult to measure at such a distance, although careful observation has indicated that it is much smaller than at first suspected, with a diameter in the region of 1,500 miles (2,400 km). In 1978 James Christy noticed that photographs of Pluto had produced an elongated image of the planet, and when earlier photographs were examined some of them showed a similar effect. Christy deduced that Pluto must have a satellite. It was given the name Charon and was found to have a diameter of around 500 miles (800 km), roughly a third of the diameter of Pluto itself. The Pluto–Charon system is more akin to a double planet than a planet and satellite, although its total mass is considerably less than that of our own Moon.

Even at perihelion, Pluto is excessively faint, and there is really very little else we can learn about this mysterious little world until we get a closer view, perhaps from a spacecraft on a future deep space mission. Until then, many of its secrets must remain hidden from us.

Observing the Asteroids and Pluto

Tables are available which show the positions of both the minor planets and Pluto, although as far as direct observational work is concerned there is very little that the amateur observer can do. In the case of Pluto a large telescope of between 10–12 inches (250–300mm) aperture is required to pick out its feeble glow as it tracks its way through the outer reaches of the Solar System. However, binoculars or a small telescope will reveal the four brightest minor planets – Ceres, Pallas, Juno and Vesta – and at least one of these should be observable at any particular time. Using tables, plot their positions on a star atlas and note the predicted change in position from night to night. Now search the area of sky in which the minor planet is located on that particular night and draw each of the stars that you can see. One of those points of light should be the asteroid you are looking for, although it will not reveal itself except through a movement across the sky. On the following night look at the area again and compare your drawing with what you see through the telescope or binoculars. If everything goes well, one of the "stars" will have changed position. Yet, as easy as this procedure sounds, things can go wrong. For example, the next few nights could be cloudy, which may mean a re-start of observations in another area of sky, or perhaps succeeding nights may be clearer and more faint stars will be visible, thereby causing confusion. Whatever the obstacles, stick with it, and you will eventually locate the asteroid you're seeking. This also applies to Pluto, although, because this planet is so faint, the search is made all that more difficult. Yet the feeling of achievement is even greater!

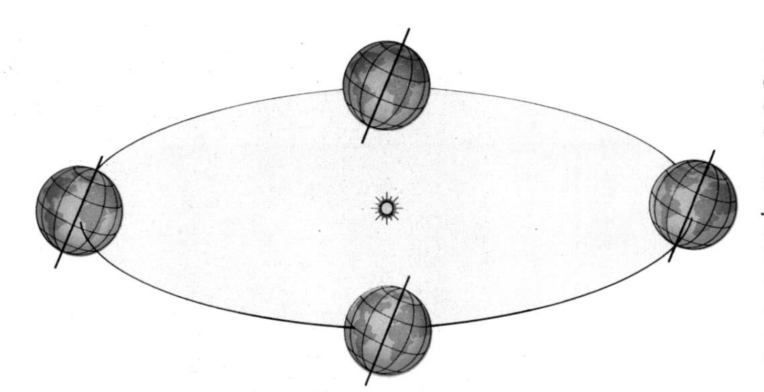

***Top left:** the tilt of the Earth's axis relative to the plane of its orbit is the main cause of the different seasons on our planet. During December the northern hemisphere is inclined away from the Sun and the southern hemisphere towards it, resulting in a northern winter and a southern summer. Six months later the situation is reversed whereby the northern hemisphere has summer and the southern hemisphere experiences winter. When any particular hemisphere is tilted towards the Sun, the Sun will be higher in the sky from locations within that hemisphere, resulting in the longer and warmer days of summer. In winter, the Sun's altitude decreases, with the result that days become shorter and colder.*
***Centre left: A:** the apparent paths of the Sun and a typical star arising from the rotation of the Earth on its axis, which is pointed towards the north celestial pole. A similar situation exists in the southern hemisphere.*
***B:** The position of a star in the sky at a particular time and from a particular location can be expressed in altitude (height in degrees above the horizon) and azimuth (its position relative to the horizon measured from north (0°) through east (90°), south (180°), west (270°) and back to north). The zenith (altitude 90°) is the point directly above the observer.*
***C:** the celestial equator is the projection of the Earth's equator on to the celestial sphere and the solstices are points on the ecliptic at which the Sun is at its maximum angular distance from the celestial equator. The points at which the ecliptic crosses the celestial equator are known as the vernal equinox (March) and autumnal equinox (September). At these times day and night are of equal length.*

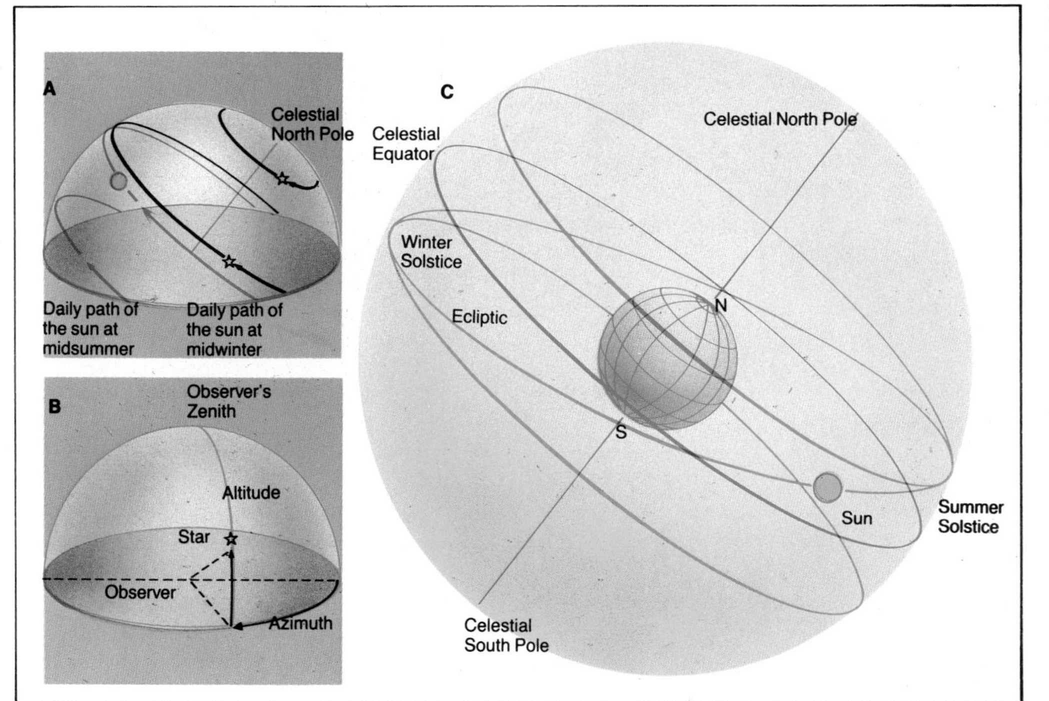

Precession: *the gravitational pull of the Sun and Moon on the Earth's equatorial bulge produces a "wobble" in the Earth's axis, similar to that of a spinning top which is slowing down. Each wobble takes 25,800 years, and over that time the celestial pole traces out a circle on the celestial sphere. The resulting change in position of the north celestial pole is shown in the diagram.*

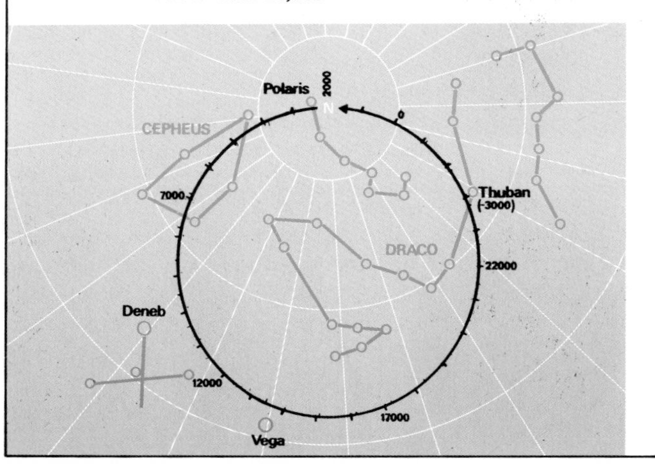

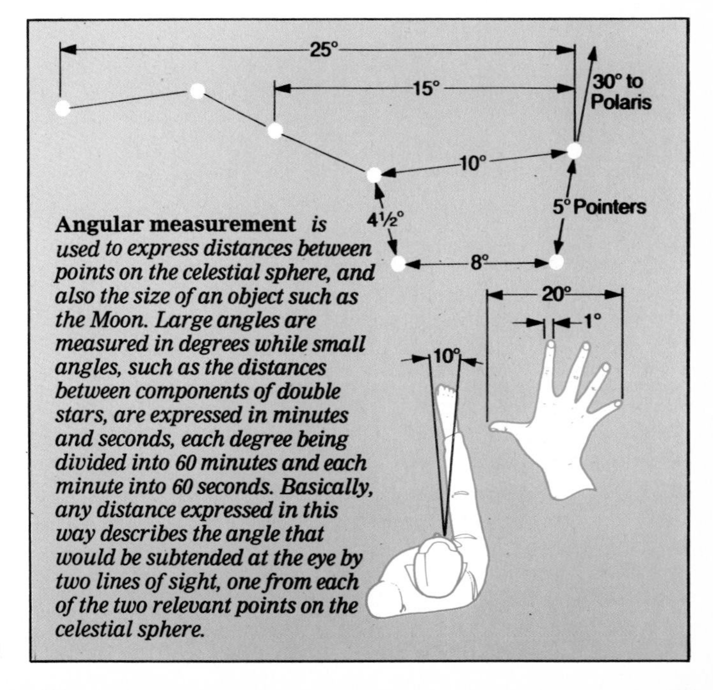

Angular measurement *is used to express distances between points on the celestial sphere, and also the size of an object such as the Moon. Large angles are measured in degrees while small angles, such as the distances between components of double stars, are expressed in minutes and seconds, each degree being divided into 60 minutes and each minute into 60 seconds. Basically, any distance expressed in this way describes the angle that would be subtended at the eye by two lines of sight, one from each of the two relevant points on the celestial sphere.*

The Sun

The Sun dominates our region of space and contains around 98 per cent of the total mass of the Solar System. The diameter of the Sun is 865,000 miles (1,392,530 km) and, although large by terrestrial standards, is small when compared with some of the other 100,000 million stars within our Galaxy. Astronomers class the Sun as a yellow dwarf and it is a typical Main Sequence star (*see* The Stars).

The outer, visible layer of the Sun is known as the photosphere and it is from here that all the Sun's radiation is emitted. The temperature of the photosphere is 11,000°F (6,000°C). However, the energy that escapes from this region is actually created at the core of the Sun, where the temperature and pressure are so immense that nuclear reactions take place whereby four hydrogen nuclei are fused together to form one nucleus of helium. During this process a tiny amount of mass is left over. This mass is released as energy and is carried by convection to the surface, where it escapes as light and heat.

If the photosphere is examined closely it will be seen to have a mottled appearance; this so-called granulation is the result of the turbulent eruption of energy at the surface. Also visible are sunspots, which appear as dark patches on the photosphere. The temperature of a typical sunspot is 7,000°F (4,000°C) and it only appears dark in comparison to the hotter, brighter region surrounding it. Sunspots have two regions – a dark, central area, or umbra, surrounded by a fainter penumbra. Sunspots are actually depressions in the photosphere, a fact which can be verified when the sun-

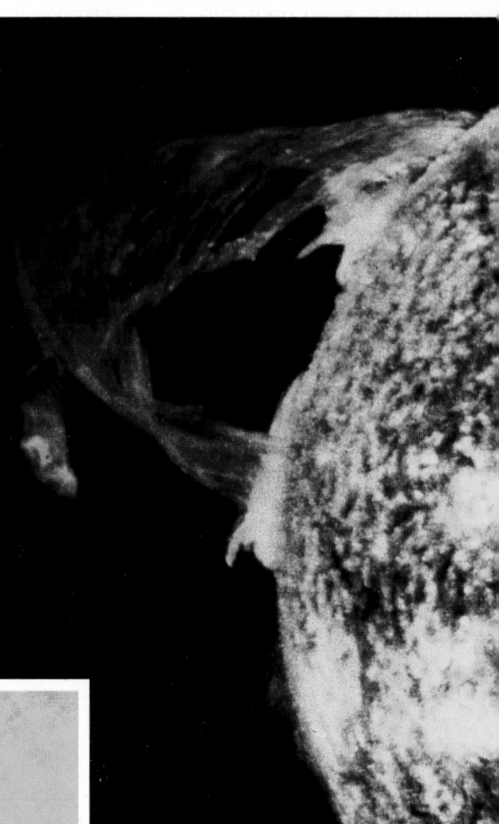

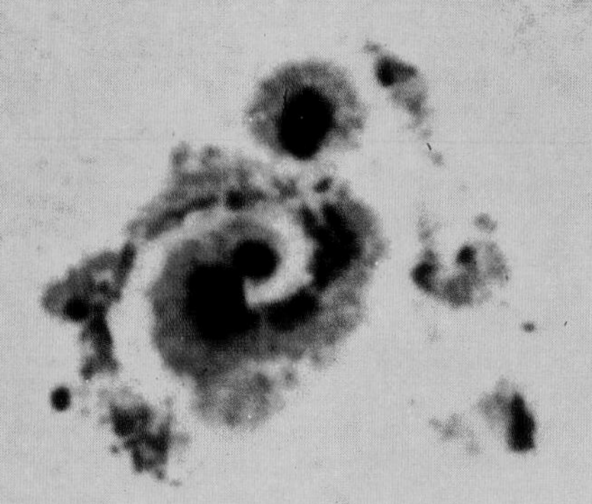

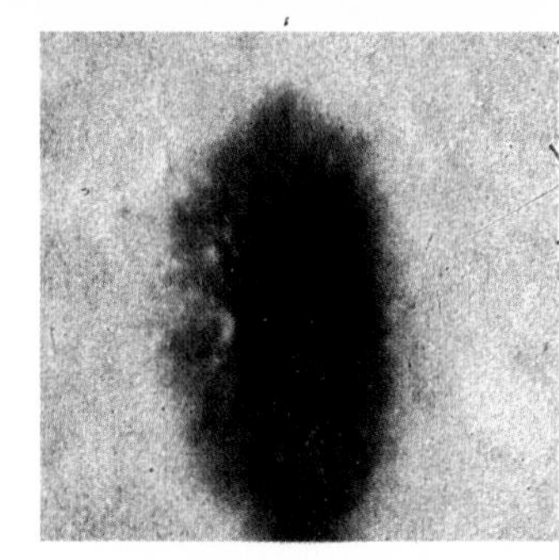

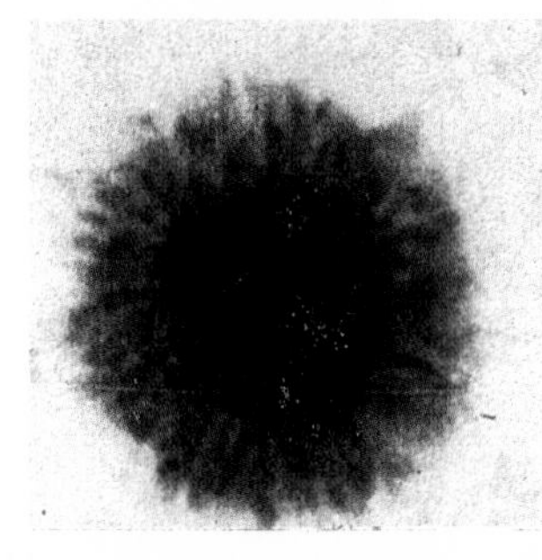

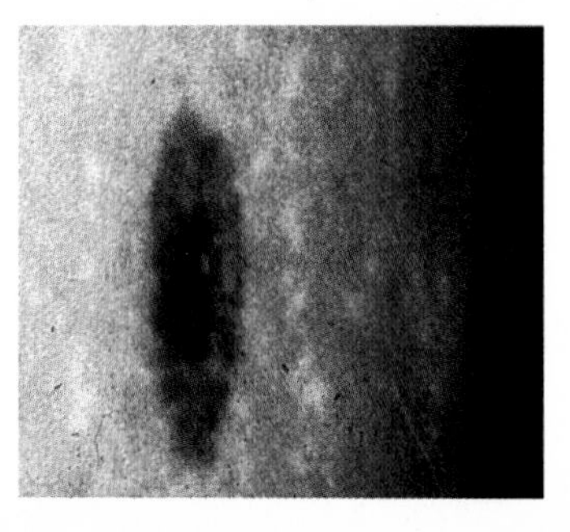

Above: *seen in December 1973, this huge solar flare was one of the most spectacular ever recorded.*
Left: *a rare spiral-shaped sunspot, photographed in February 1982.*
Far left: *cross-sectional view of the Sun showing the hot, central core where solar energy is produced.*
Left and below: *as a sunspot approaches the solar limb the penumbra on its far side is seen to be wider than that on the near side. This is the Wilson Effect and it indicates that sunspots are actually depressions.*
Right: *section of the Sun's outer layers showing the wide range of temperatures in this region.*

spot is seen near the solar limb. When observed in this region there is a marked foreshortening and the penumbra on the side of the sunpot nearest us appears narrower. This phenomenon is known as the Wilson Effect and was first observed by Alexander Wilson in 1769. Although sunspots can appear at any time, there is a well-defined sunspot cycle of around 11 years' duration in which there is a regular increase and decrease in the level of sunspot activity.

The chromosphere lies above the photosphere and it is through here that solar energy passes on its way out from the Sun. It is in this region that faculae and flares are seen. Faculae are bright luminous clouds, comprised mainly of hydrogen, which are observed above the regions where sunspots are about to form. Sunspots also give rise to flares, which are bright filaments of hot gas seen to emerge from sunspot regions. Flares can produce vast increases in the output of charged particles from the Sun, which in turn may produce a rise in the frequency and intensity of auroral displays here on Earth.

Beyond the chromosphere lies the corona, which forms the outer part of the Sun's "atmosphere". It is within the inner part of the corona that prominences appear – vast clouds of glowing gas that erupt from the upper chromosphere – and which are observed on the limb of the Sun. There are two basic forms of prominence: quiescent prominences are fairly stable and may persist for quite a few months before they disperse, while eruptive prominences which are much more active and can display marked changes in shape from minute to minute. The outer region of the corona stretches out into interplanetary space and consists of particles that are travelling relatively slowly away from the Sun.

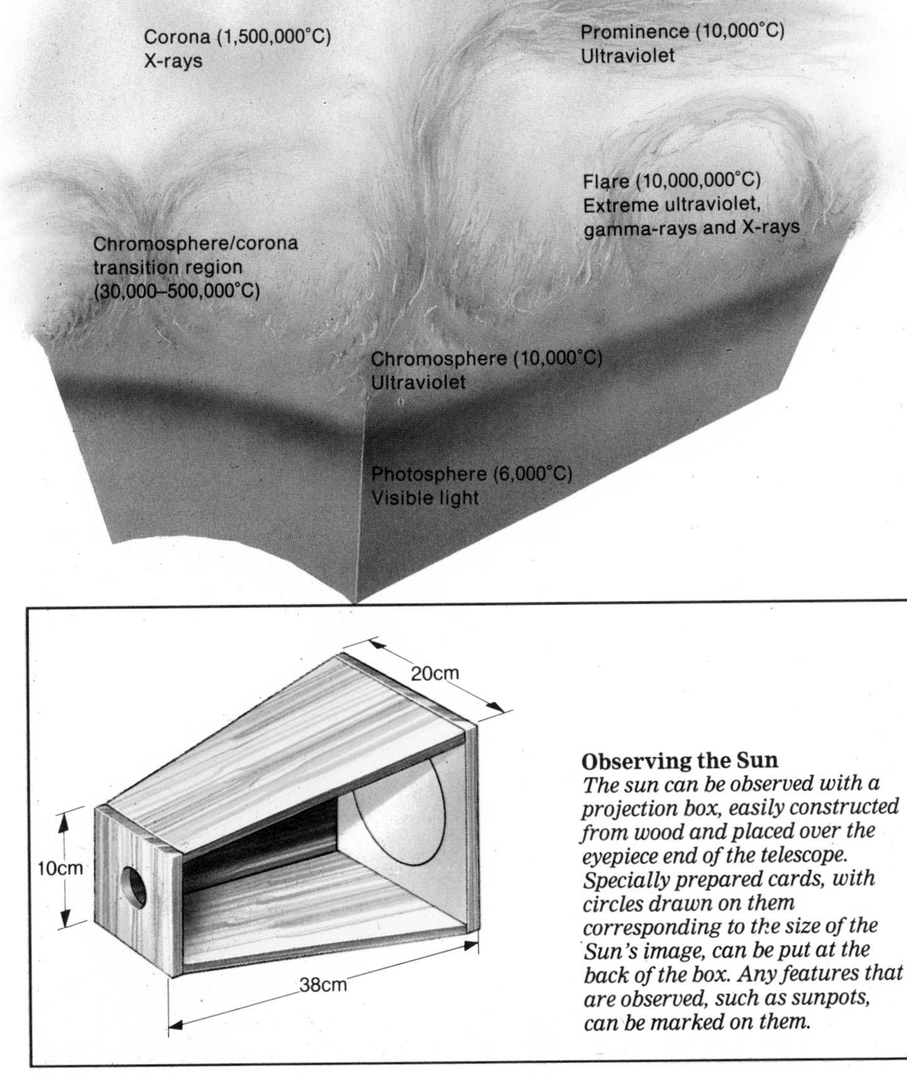

Observing the Sun
The sun can be observed with a projection box, easily constructed from wood and placed over the eyepiece end of the telescope. Specially prepared cards, with circles drawn on them corresponding to the size of the Sun's image, can be put at the back of the box. Any features that are observed, such as sunpots, can be marked on them.

Observing the Sun

The Sun is by far the most visually accessible object in the sky. Unfortunately, as far as the amateur is concerned, many of the features associated with the Sun are not observable unless special equipment is used. However, sunspots, faculae and granulation are within reach of the amateur.

One thing that must be borne in mind is that the Sun must never be observed directly, as permanent eye damage could easily be caused.

The only safe way to observe the Sun is by the projection method, whereby the telescope is used to project an image of the Sun onto a white screen. Solar projection screens are supplied with some new telescopes, although a piece of card will suffice. The projection method can be carried out with either reflectors or refractors, although the design of refracting telescopes makes them ideal for the purpose. In order to keep direct sunlight away from the screen a piece of card should be placed around the telescope tube. This will cast a shadow onto the screen and a brighter, clearer image of the Sun will be seen. Another way to overcome the effects of stray light is to construct a projection box out of wood with one side left off. This provides an enclosed area around the screen from which most daylight is excluded and gives an unobstructed view of the solar image.

For most solar observation an image of around 6 inches (15 cm) is suitable. A good view of the solar disc can be obtained with a low magnification of around 25× or 30×, although increasing this to between 100× and 120× may bring out details of granulation on the photosphere. Sunspots are far easier to see and it is interesting to make a daily record of sunspot activity. Provided that a good, clear image is obtained, faculae may also be seen near the limb and when they are, keep an eye on the area in question to see if sunspots appear.

The Moon

The Moon is our closest celestial neighbour and has always held a fascination for mankind. It illuminates our night sky and the constantly changing lunar phases have been a source of wonder throughout the ages. It also has a physical effect on our planet in that its gravitational influence produces the tides which sweep around our oceans. (The Sun also plays a part in tidal production, although to a much lesser extent.)

Although the Moon is normally classed as a satellite of the Earth, its diameter of 2,160 miles (3,476 km) makes it over a quarter the diameter of our planet. It may, therefore, be better to regard the Earth and Moon as a double planet rather than a planet and satellite. The Moon orbits the Earth in a period of 27.3 days, which is exactly equal to its period of axial rotation. As a result, the Moon has a "captured rotation", which means that it keeps the same face turned towards the Earth. However, the distance of the Moon varies, ranging from 221,468 miles (356,410 km) at its closest point (perigee) to 252,716 miles (406,697 km) at its furthest point (apogee). Because of this the Moon's orbital speed alters slightly, and is greater at perigee than apogee. This results in the orbital and axial rotation periods becoming slightly out of step and causes the Moon to "wobble" slightly. This process is known as "libration" and because of it we are able to see up to 59 per cent of the total lunar surface.

The Origin of the Moon

Opinions vary as to the origin of the Moon. It was originally believed that the Moon was once part of the Earth and that long ago, shortly after the Earth's formation, a chunk of our then rapidly spinning planet broke away to form the Moon. This theory has now been discarded, as has the similar idea that smaller particles broke away from the Earth and collected together to form our satellite. It has also been suggested that the Earth and Moon were formed at roughly the same time, the Moon coming into being as the result of the collecting together of dust and debris in orbit around the Earth. However, it now seems that both the Earth and the Moon formed in different regions of the Solar System

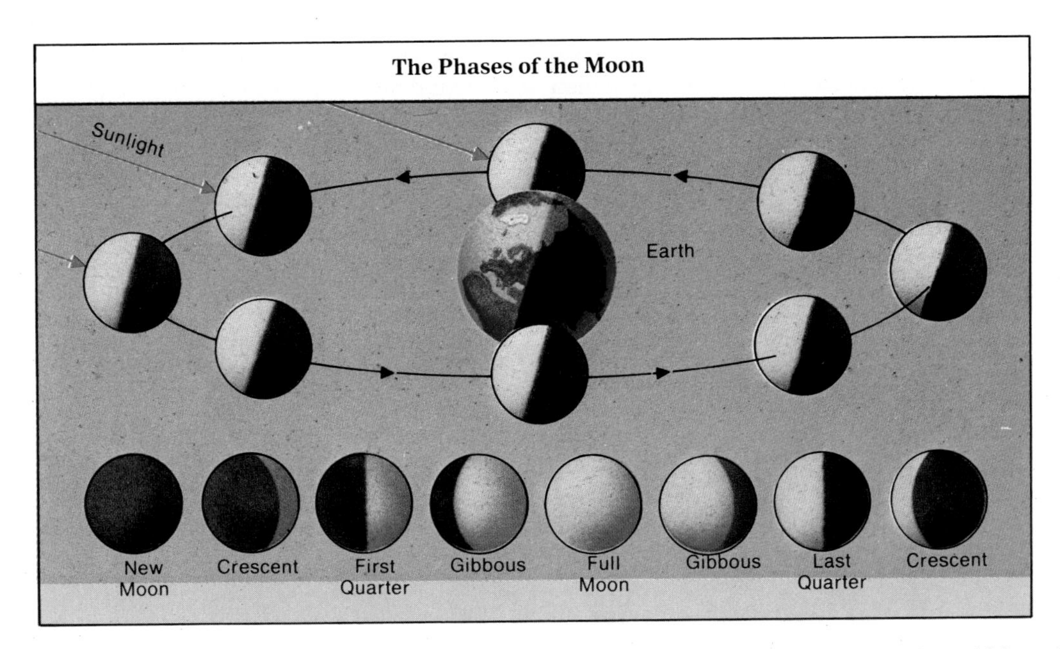

As the Moon orbits the Earth it appears to change shape through a regular cycle of phases. When the Moon is situated between the Earth and Sun its dark side is facing us and it is invisible. However, as the Moon travels around the Earth its illuminated half gradually reveals itself, and our satellite is seen to wax through crescent, first quarter and gibbous phases. Eventually the point is reached where the Moon is opposite the Sun in the sky and the fully lit half is turned towards us. It is then that we see a Full Moon. After this the sequence of phases is reversed, the cycle finally being completed when New Moon occurs again.

and came together as a result of the Moon being captured by the Earth's gravity as it wandered close to our planet. Yet even this theory has its drawbacks, and it may be a long time yet before we can fully explain the origin of our only natural satellite.

The Lunar Surface

Although the origin of the Moon remains something of a mystery, the subsequent evolution of the lunar surface is far better understood. Even a casual glance will reveal light and dark areas on the Moon, and closer examination will show the true nature of these regions, together with examples of many different features, including rugged craters and basins, expansive maria, deep valleys and lofty mountain ranges.

Shortly after the Moon formed it began to cool down, its outer layers solidifying to form a surface crust. During this period the Solar System contained much interplanetary debris and the lunar surface was subjected to a great deal of meteoritic bombardment, which led to the formation of vast numbers of craters. Volcanic activity then followed and filled the low-lying areas with lava, creating the mare-type terrain. A further period of reduced meteoritic activity resulted in a scattering of craters on the maria. This left the Moon as we see it today, with the bright, cratered highlands, which cover roughly 75 per cent of the lunar surface, contrasting sharply with the darker maria.

Observing the Moon

As the Moon passes through its cycle of phases the terminator (the boundary between the light and dark hemispheres) is seen to cross the lunar disc. Lunar features are best seen when near the terminator, as the craters, mountains and valleys will cast dark shadows, thereby highlighting the terrain. Identification of many notable features can be carried out using binoculars and can begin when the Moon is only two days old. It will now be visible as a thin crescent low in the evening sky after the Sun has set. The eastern section of Mare Crisium will now be in view, as will the conspicuous crater Langrenus to the south. The walls of this crater are over a mile high and stand out extremely well under these conditions. Craters Vendelinus and Petavius may also be seen further to the south.

When the Moon is five days old many of the dark maria stand out quite well, including Mare Serenitatis, Mare Tranquillitatis, Mare Nubium and Mare Fecunditatis. Mare Nectaris is also well placed, as is the magnificent crater Theophilus on its north-west side. Theophilus is around five miles deep and has a large central peak, which is fairly prominent.

At first quarter, we can see only a quarter of the total lunar surface. Although half of the total surface is illuminated, we can only see half of that, the other portion being behind the body of the Moon as seen from Earth. Many features are now seen to good advantage. The crater Hipparchus, normally difficult to see, stands out quite well just to the south of the lunar equator. Mare Serenitatis is bordered on its western side by the Apennine and Caucasus mountain ranges, while further to the north the Alpine Valley cuts across the rugged terrain between Mare Imbrium (still in darkness) and Mare Frigoris. This region is dominated by the Alps, a mountain range whose loftiest peaks reach a height in excess of 6,560 feet (2,000 m) above the surrounding plains. The Alpine Valley may be difficult to see through binoculars, but is well worth the search.

Shortly after first quarter some magnificent crater chains come into view. Ptolemaeus, Alphonsus, Alpetragius and Arzachel can be seen stretching down from just below the lunar equator, while Pilatus and Tycho are easily distinguished further to the south. Named after the famous Danish astronomer Tycho Brahe, this crater has a diameter of 53 miles (85 km) and a depth of around 3 miles (5 km).

As the Moon moves towards full the rest of the nearside maria come into view, including Mare Imbrium and Mare Nubium. Sinus Iridum (Bay of Rainbows), high in the north-west, begins to become prominent after ten days, as does the magnificent crater Copernicus. This crater has a diameter of 58 miles (93 km) and its terraced walls rise to almost $2\frac{1}{2}$ miles (4 km) above the crater floor. Copernicus can be seen quite well at any phase, although it is particularly prominent when near the terminator. On the northern reaches of Mare Imbrium can be seen Pico, a lofty mountain some $1\frac{1}{2}$ miles (2.5 km) high, which casts its shadow over the surrounding flat landscape. Just to the north of Pico is the crater Plato, a walled plain around 60 miles (100 km) across.

Gradually the crater Kepler comes into view to the west of Copernicus, in the central regions of Oceanus Procellarum. To its south can be seen Mare Humorum, with Gassendi, a large ringed plain, on its northern edge.

During Full Moon the lunar disc is virtually free of shadow and there is little relief to highlight the lunar features. However, it is now that the ray systems associated with a number of craters become prominent. Copernicus, Kepler and Tycho are among those craters that are seen to have bright streaks extending radially from them. For a long time their origin remained a mystery, although it is now known that they are the result of the ejection of fine material which took place as the craters were formed by meteoritic impact.

The Moon's surface bears testament to its violent history, with the craters and maria coming into being as a result of a mixture of meteoritic bombardment and vulcanism. A lack of atmosphere around our satellite means that there are no erosive elements that can destroy these features in the way that so many geological features have falley prey to the weather systems on Earth. There is a wealth of detail observable and it can be very enjoyable to scan the lunar surface picking out the individual features. As with all types of observation some record should be kept, if only to conjure up memories of your nights at the telescope. Sketching the features that are seen is a marvellous way of keeping these memories, and with a little practice quite good drawings can be produced.

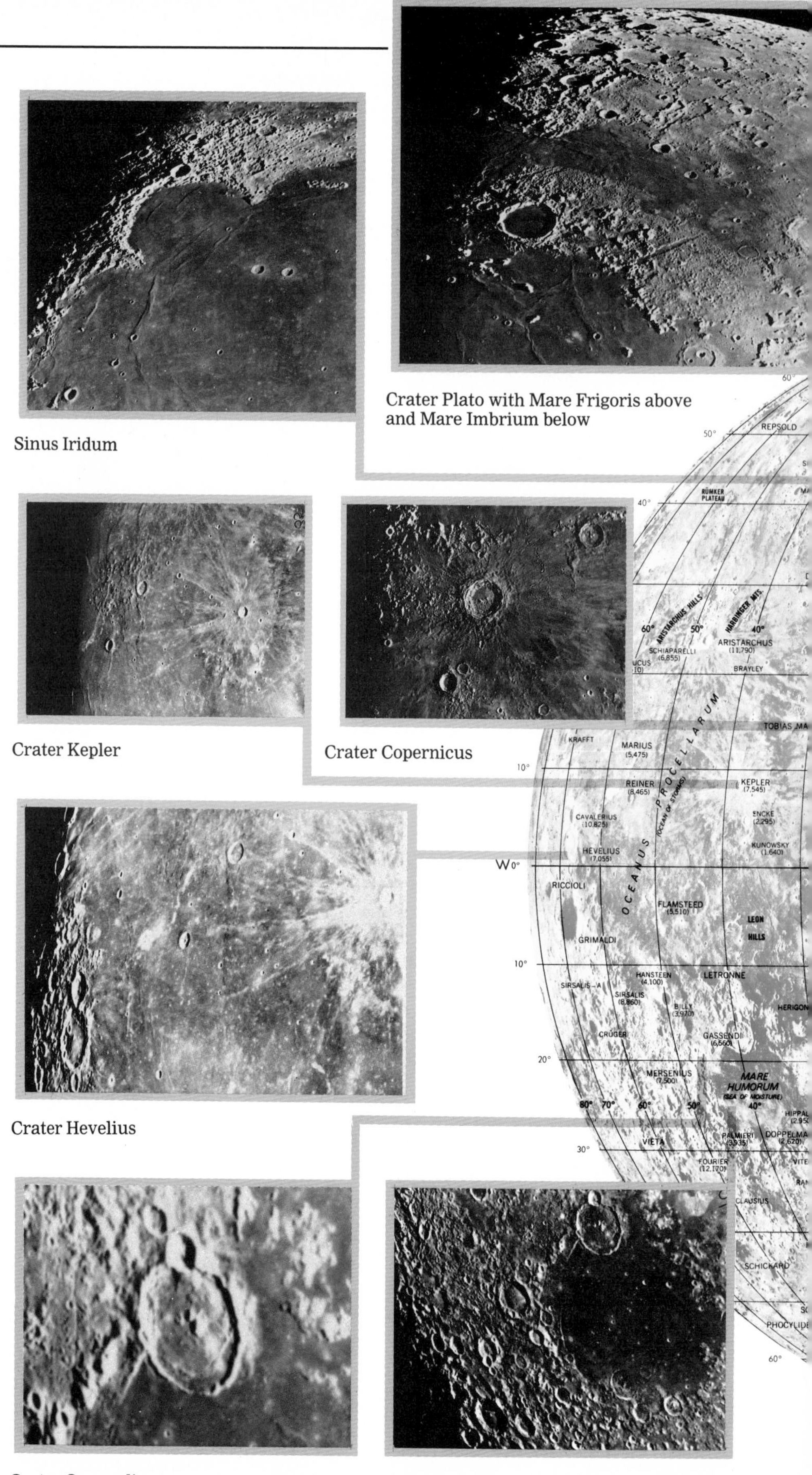

Crater Plato with Mare Frigoris above and Mare Imbrium below

Sinus Iridum

Crater Kepler

Crater Copernicus

Crater Hevelius

Crater Gassendi

Mare Humorum

Drawing of lunar craters Atlas and Hercules made on 29 August, 1983, by Peter Grego using a 60 mm refractor at 100× magnification. This shows what can be achieved with basic equipment and a little practice.

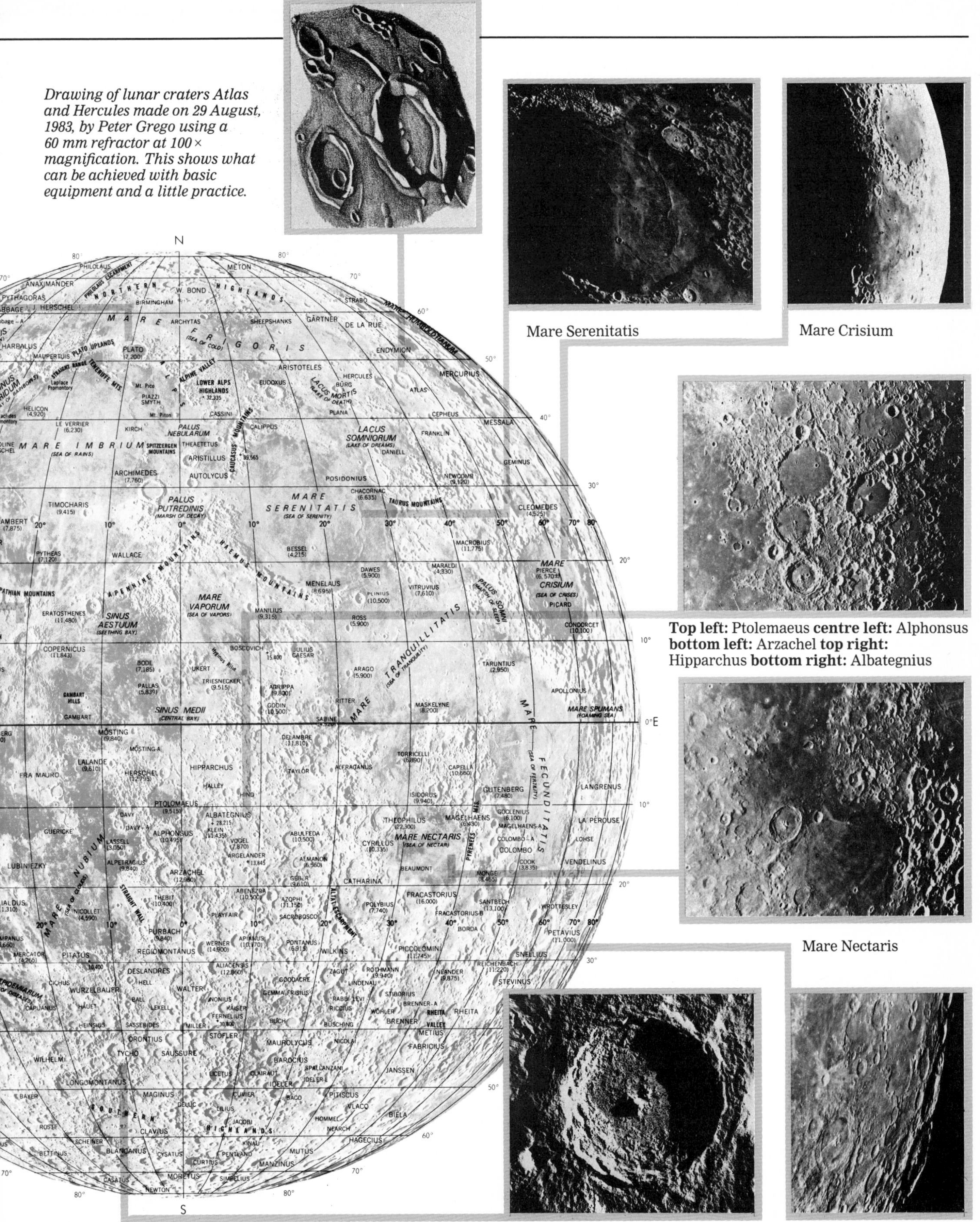

Mare Serenitatis

Mare Crisium

Top left: Ptolemaeus **centre left:** Alphonsus **bottom left:** Arzachel **top right:** Hipparchus **bottom right:** Albategnius

Mare Nectaris

Crater Tycho

Crater Petavius

The Inner Planets

The Sun's family of planets is split into two quite distinct groups: the four inner, terrestrial planets, Mercury, Venus, Earth and Mars – comparatively small and rocky – are similar in many ways to each other. As a group, however, they contrast sharply with the four outer gas giants, Jupiter, Saturn, Uranus and Neptune. All four inner planets bear the scars of meteoritic impacts on their surfaces, and vulcanism also seems to have played a major role in the shaping of these worlds.

Closest to the Sun is Mercury, the smallest of the terrestrial planets. Mercury has a diameter of 3,030 miles (4,870 km) and orbits the Sun once every 88 days at a mean distance of 35,984,590 miles (57,910,000 km). However, its orbit is quite eccentric, and the distance of Mercury from the Sun varies between 28,521,700 miles (45,900,000 km) at its closest point (perihelion) to 43,310,750 miles (69,700,000 km) when at its furthest point (aphelion). Because Mercury is the innermost of the planets it is never seen against a truly dark background from Earth. As a result early observations of the planet were unreliable.

Our knowledge of the Mercurian surface improved greatly with the American Mariner 10 fly-by mission. Photographs sent back during 1974 and 1975 revealed a heavily cratered world. Also in evidence were mountainous regions, valleys, scarps, ridges and plains or basins, the largest feature being the Caloris Basin, an enormous plain 800 miles (1,300 km) in diameter.

The next planet out from the Sun is Venus, one of the brightest objects in the sky. Venus has a diameter of 7,520 miles (12,104 km) and orbits the Sun at a mean distance of 67,234,200 miles (108,200,000 km). The brilliance of Venus is due to sunlight being reflected from dense clouds which completely cover the planet, and which have been found to contain large amounts of sulphuric acid. They hide a hostile surface where the temperature is in the order of 930°F (500°C) and the atmospheric pressure is around 90 times that of the Earth at sea level.

The Venusian clouds prevent astronomers from seeing the surface, but radar mapping by space missions has shown a largely flat terrain, although there are some notable highland areas, including Aphrodite Terra and Ishtar Terra. The latter supports a number of mountains, including Freyja and Maxwell, both of which tower several miles above the surface of the planet. Other mountainous regions are also in evidence, as are craters and valleys, including Diana Chasma, one of the largest features of its kind in the Solar

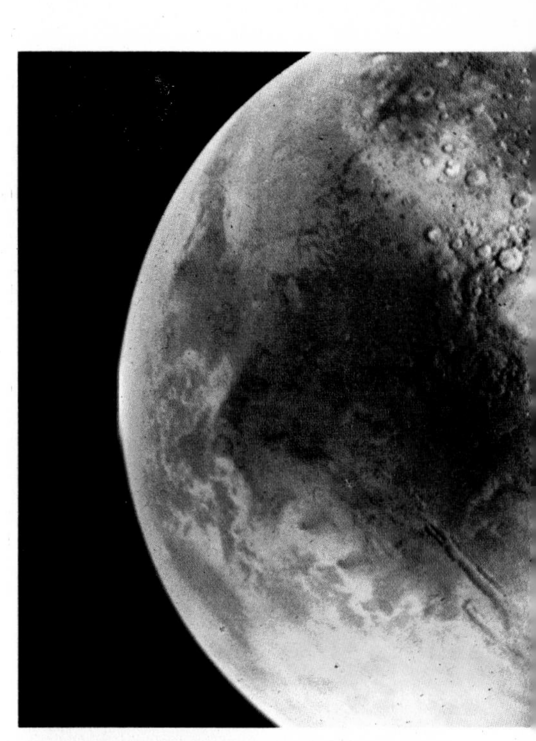

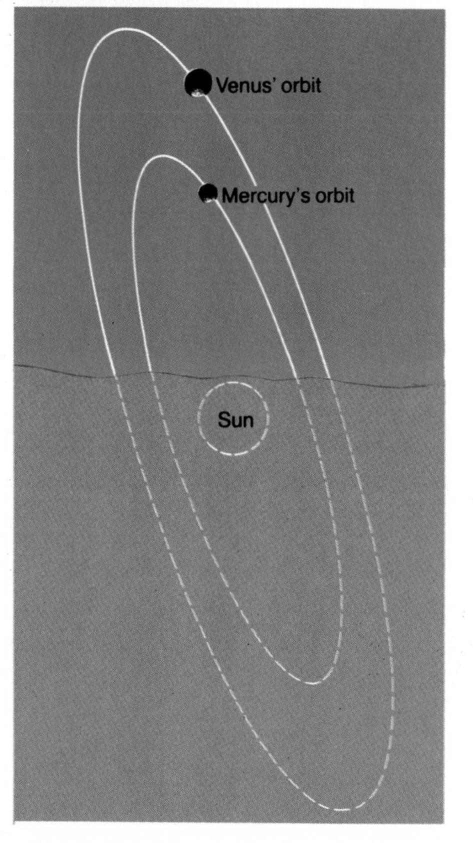

Top: *Mars photographed by the Viking 1 orbiter in June 1976.*
Above: *Mercury and Venus always appear close to the Sun because their orbits lie inside that of the Earth.*
Left: *Venus and the crescent Moon in the evening sky.*

RETROGRADE MOTION

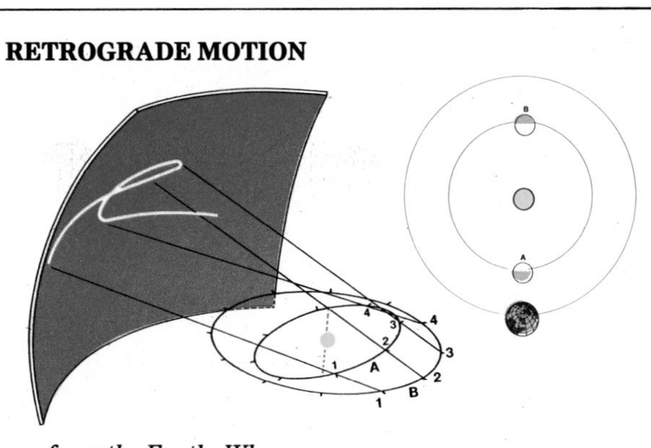

Right: *the apparent path of Venus as seen from Earth. The corresponding positions of Venus (**A**) and Earth (**B**) are indicated numerically. The path of Mercury is similar although its motion through the sky is more rapid.*
Far right: *an inferior planet (either Mercury or Venus) is at inferior conjunction (**A**) when between the Earth and Sun and at superior conjunction (**B**) when on the opposite side of the Sun from the Earth. When a planet is at conjunction it is difficult to observe because of its proximity to the Sun.*

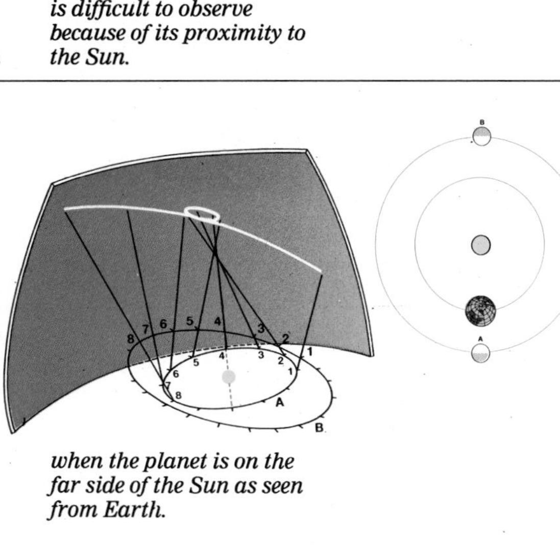

Right: *at around the time of opposition a superior planet appears to undergo retrograde motion as the higher orbital speed of the Earth causes it to "overtake" the planet. The corresponding positions of Earth (**A**) and the superior planet (**B**) are shown numerically.*
Far right: *a superior planet is at opposition (**A**) when it is opposite to the Sun in the sky and is consequently well placed for observation. Superior conjunction (**B**) occurs when the planet is on the far side of the Sun as seen from Earth.*

System.

The Earth is unique in that almost three-quarters of its surface is covered by water. It is the largest of the terrestrial planets with a diameter of 7,930 miles (12,756 km) and its surface shows evidence of its turbulent geological past. There are a great number of meteorite craters, although they are somewhat difficult to detect, and vulcanism has figured prominently in the formation of the oceans and, to a large extent, in the composition of the atmosphere that surrounds our planet. The distribution of land masses is now known to be the result of "continental drift", a process whereby sections of the Earth's crust form plates which slowly drift about.

The outermost of the terrestrial planets is Mars. Like Mercury, it has an eccentric orbit, its distance from the Sun varying from 128,005,000 miles (206,000,000 km) at perihelion to 154,725,000 miles (249,000,000 km) at aphelion. Mars has been the destination for a number of space probes, and pictures sent back have revealed many interesting features, including craters, canyons and volcanoes. Mars can boast both the largest volcano and the greatest canyon in the Solar System. Olympus Mons is a gigantic volcano some 16 miles (25 km) high and 340 miles (550 km) across its base. Valles Marineris is a huge canyon which runs through the equatorial region of Mars. It has a length of 2,480 miles (4,000 km) and is over 44 miles (70 km) across at its widest point.

There are two satellites in orbit around Mars, both of which were discovered by Asaph Hall in 1877. Named Phobos and Deimos they are very small and are probably nothing more than asteroids that have been captured by the Martian gravity.

Observing the Inner Planets

Mercury has a reputation as being difficult to see, but it can be found quite easily with binoculars or even the naked eye. The secret is to look for the planet when it is at its maximum elongation, or greatest angular distance from the Sun. If the sky is reasonably clear Mercury may be glimpsed with the unaided eye, hovering at the edge of the twilight glow. If binoculars are needed, sweep the area until you find the bright starlike object. Once found, Mercury can be followed from day to day as it quickly moves in towards the Sun. The best that can be expected with small telescopes is to catch sight of the phase, although this is possible only in ideal observing conditions.

Venus is so bright that it can be seen in broad daylight. To do this a telescope with setting circles must be used as casual sweeping of the sky may bring the Sun into view, with disastrous results to the eye. Venus is best observed when the sky is not completely dark, as otherwise the glare from the planet will be too intense. Like Mercury, Venus displays phases observable with good binoculars or a small telescope. If you follow the planet as its angular distance from the Sun alters, the phase will be seen to change as it moves towards or away from the Sun.

Mars is also something of a disappointment for the observer, although large telescopes may show one or two dark features on its surface together with a polar ice cap that will be seen to expand and contract as the Martian seasons change. Some clouds may also be seen suspended high above the Martian surface and occasionally dust storms will blot out some, or even all, surface features. The best time to observe Mars is when the planet is at opposition, as it is then that it will be at its closest point to Earth. It is enjoyable to plot Mars as it moves through the sky, particularly near opposition, when it will be seen to undergo its looping retrograde motion as the Earth "overtakes" it.

The Outer Planets

Jupiter is the largest planet in the Solar System with an equatorial diameter of 88,240 miles (142,000 km), over eleven times that of the Earth. However, it has a very short axial rotation period of just over 9h 50m, which has resulted in a marked polar flattening and a polar diameter of only 83,390 miles (134,200 km). Jupiter is composed mainly of hydrogen, which makes up over 80 per cent of the outer, visible cloud layer. Helium is also present here, together with traces of other elements. Telescopically, this outer layer is seen to be divided into a number of dark belts and bright zones.

Jupiter has 16 satellites, the four largest being Io, Europa, Ganymede and Callisto. A ring system has also been found around Jupiter, but it is excessively faint and totally invisible from Earth.

The second largest planet is Saturn, with an equatorial diameter of 74,130 miles (119,300 km). Saturn's rotation period is a little over ten hours, which has resulted in a noticeable polar flattening. Saturn is made up largely of hydrogen, together with small amounts of helium and methane, and its overall density is only around 70 per cent that of water.

Saturn has the largest number of satellites – more than 20 – in the Solar System, the largest and brightest of which is Titan, with its dense atmosphere made up almost entirely of nitrogen but with significant amounts of helium present.

Saturn's most notable feature is, without doubt, its ring system, which has led to Saturn being described as one of the most beautiful objects in the sky. The rings are split into a number of different parts, including the bright A and B Rings and a fainter C or Crepe Ring. There are various gaps in the ring system, the most prominent of which is the Cassini Division, discovered by Giovanni Cassini in 1675 and which separates the A and B Rings. Ring A is split by the Encke Division, discovered by Johann Encke in 1837. Space probes have shown that the main rings are actually made up of large numbers of narrow ringlets, giving an overall impression of a gigantic gramophone record.

The seventh planet out from the Sun is Uranus, discovered by William Herschel in 1781. Uranus has an equatorial diameter of 32,190 miles (51,800 km) and orbits the Sun once every 84.01 years at a mean distance of 1,783,135,000 miles (2,869,600,000 km). In 1977, a system of five rings was found around the planet. Since then, partly through Earth-based observation and partly through the analysis of photographs taken by the American Voyager fly-by mission in early 1986, this number has been increased to eleven. All these rings are actually bright regions of a single, tenuous ring that girdles the planet.

Uranus has at least 15 moons. The two largest – Titania and Oberon – were discovered by William Herschel in 1787.

Neptune is the outermost of the gas giants. It is slightly smaller than Uranus with an equatorial diameter of 30,760 miles (49,500 km). No ring system has yet been detected around Neptune, although the Voyager spacecraft, due to pass the planet in 1989, may show otherwise. Two satellites are known to be in orbit around Neptune; Triton and Nereid, and a third is suspected.

__Left:__ Jupiter with its four brightest satellites, as seen through a small telescope.
__Below:__ photograph of Jupiter from Voyager 1, taken in January 1979, showing the Red Spot, a region of high pressure in the Jovian atmosphere.

Observing the Outer Planets

Jupiter is the easiest and most rewarding planet to observe either with binoculars or a small telescope. It is the largest planet in the Solar System and as such will show a disc even through binoculars. A small telescope will show Jupiter's polar flattening, the equatorial regions seeming to bulge out, giving the planet a somewhat

***Right:** as the relative positions of Earth and Saturn change, the angle at which Saturn's ring system is presented alters.*
***Below:** different views of Saturn and its rings.*
***Below right:** plan of Saturn's ring system.*

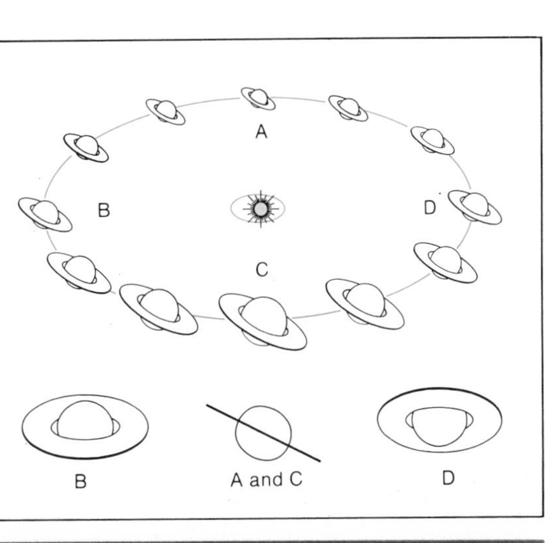

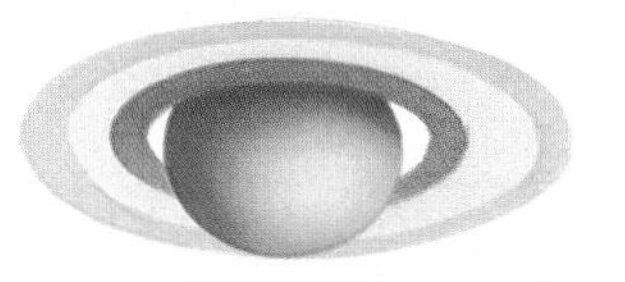

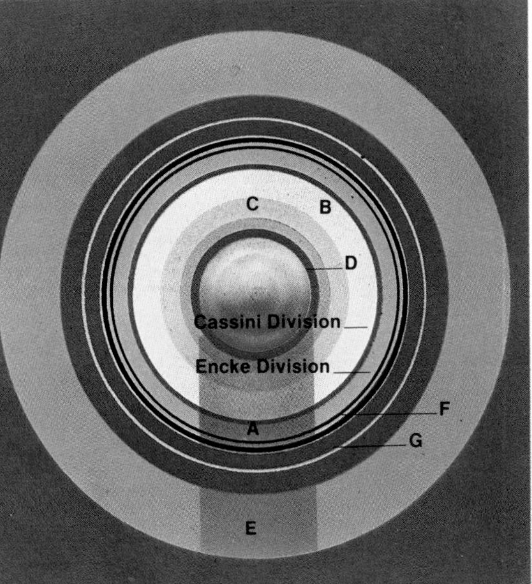

oblate appearance. Some of the more prominent cloud belts will also be seen, particularly those lying close to the Jovian equator.

It is fun to watch the four brightest satellites as they shift position from night to night as they orbit the planet. Try sketching what you see and build up a record of their movements.

To see the rings of Saturn a telescope of at least $2\frac{1}{4}$ inches (60 mm) diameter will be needed. The inclination of Saturn's rings with respect to Earth alters, and at certain times we see either the north or south face presented to us, while at other times the rings are edge on and extremely difficult to see. When the rings are well presented, a telescope of 4 inches (100 mm) aperture should reveal the Cassini Division under good seeing conditions.

Of Saturn's satellites, only Titan is bright enough to be seen with either binoculars or a small telescope, although larger instruments of 4 inches aperture or more should bring out Rhea and possibly Tethys and Dione.

Although Uranus is just visible to the unaided eye under ideal conditions, a telescope of at least $2\frac{1}{4}$ inches aperture, together with a magnification of 50× or more, will be needed to resolve the planet into a disc. If it is observed over a period of two or three days its motion against the starry background should be quite noticeable.

Neptune is fainter than Uranus and binoculars will be needed to help identify the planet. Its position can also be plotted on a star atlas that shows stars down to at least 8th or 9th magnitude and its movement against the background of stars will be readily seen only after several nights. A telescope of at least 4 inches aperture and a magnification of 100× is necessary to show Neptune as a disc.

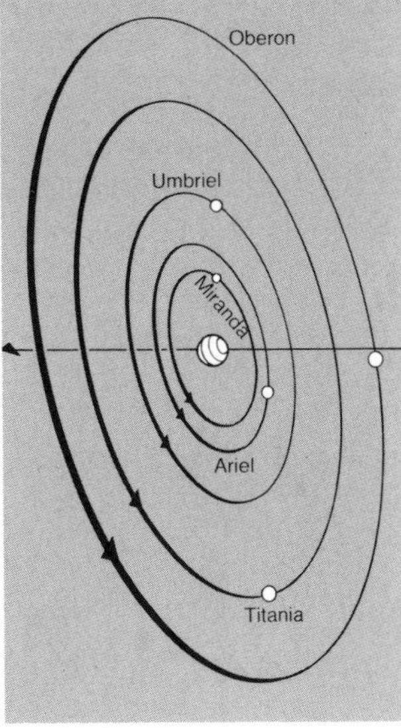

***Left:** orbits of Uranus' five brightest satellites.*
***Far left:** photograph taken by Voyager 2 in July 1985 showing Uranus with Ariel (above Uranus), Umbriel (below and to the right), Titania (at far right, beyond Umbriel) and Oberon (far left).*

Comets, Meteors and Meteorites

Comets

Beyond the orbit of Pluto, and stretching out to a distance of around 2 light years from the Sun, astronomers believe that there is a vast cloud or shell of primeval matter completely surrounding the Solar System, and it is from here that comets are thought to originate. Occasionally, the gravitational influence of a passing star may create slight but significant disturbances in this cloud, and clumps of material may be released and sent on the long journey down towards the inner reaches of the Solar System. After a journey of perhaps thousands of years, the light and heat from our parent star begins to vaporize ice locked within this clump. Eventually, as the comet travels in towards the Sun its speed increases, and the energy from the Sun acts on the cloud, or coma, forcing material away to form one or more tails, which are comprised either of gas or dust. As the comet rounds the Sun and begins to move away, the tails and coma gradually disappear as the effects of solar energy decrease. Eventually, all that remains is the icy nucleus continuing its journey around the Solar System.

Although comets can appear awesome and impressive, they are not what they seem. A typical comet has very little mass and is comprised of very tenuous material.

As a comet orbits the Sun, each close approach to our star causes it to lose some of its mass, and for those comets with very short periods this mass loss can have disastrous effects. On the other hand, comets with longer periods, such as Halley's (76 years) and Grigg-Mellish (164.3 years) remain much more impressive.

Meteors and Meteorites

There are also countless numbers of tiny particles in orbit around the Sun, and every so often one of these particles may enter the Earth's atmo-

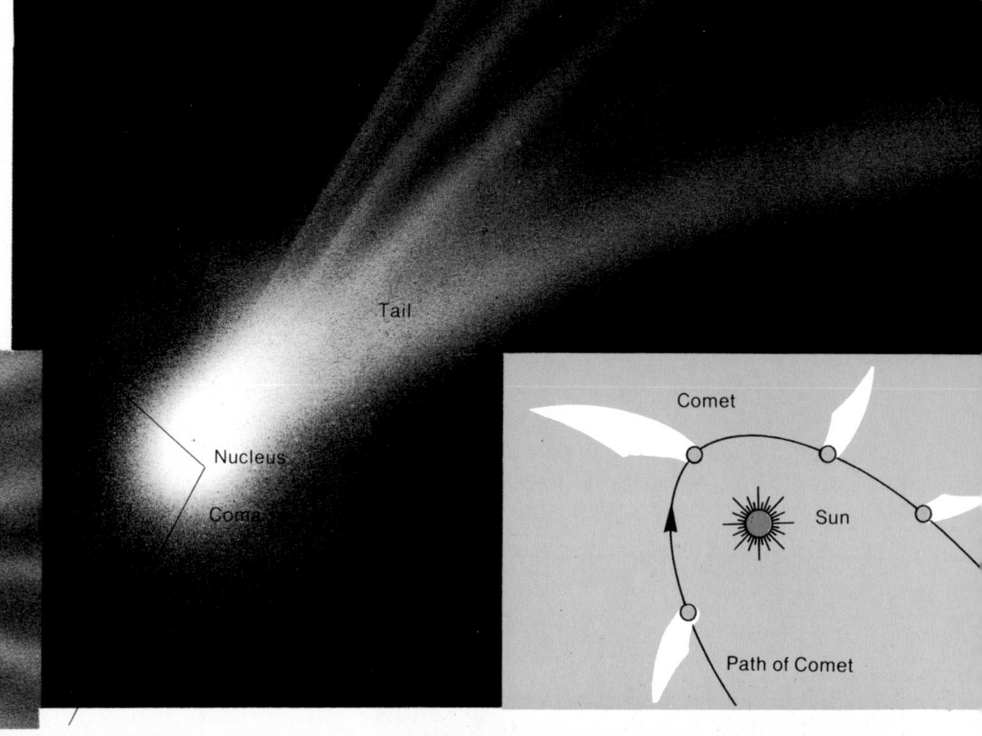

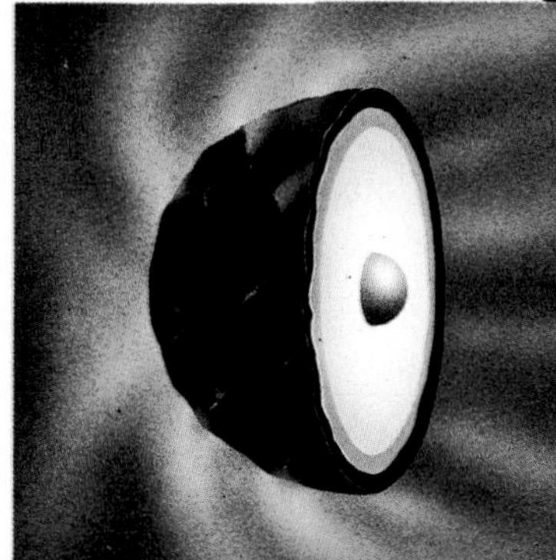

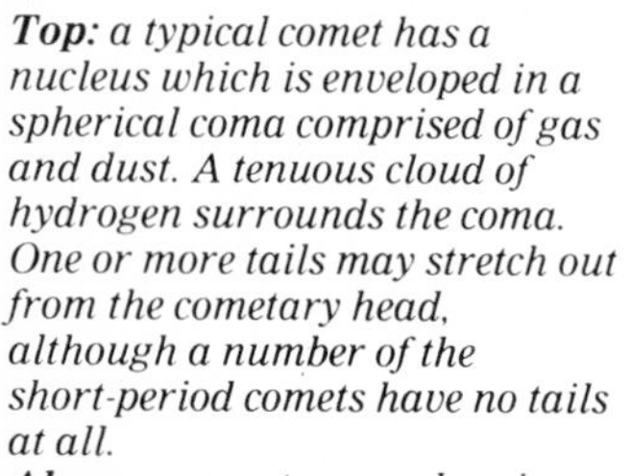
Top: *a typical comet has a nucleus which is enveloped in a spherical coma comprised of gas and dust. A tenuous cloud of hydrogen surrounds the coma. One or more tails may stretch out from the cometary head, although a number of the short-period comets have no tails at all.*
Above: *a cometary nucleus is covered with a dark crust containing vents through which gas and dust is emitted from the interior.*
Above right: *the solar wind forces a comet's tail to point away from the Sun.*
Right: *Halley's Comet on its last return in 1985-86.*

Meteor shower	Date of return	Maximum	Location	Comment
Quadrantids	1-6 Jan.	4 Jan.	Boo.	Quite fast, blue
Corona Australids	14-18 Mar.	16 Mar.	CrA.	
Lyrids (April)	19-24 Apr.	21 Apr.	Lyr.	Fast, brilliant
Aquarids	1-8 May	5 May	Aqr.	Fast, persistent
Lyrids (June)	10-21 Jun.	15 Jun.	Lyr.	Blue
Ophiuchids	17-26 Jun.	20 Jun.	Oph.	
Capricornids	10 Jul.-15 Aug.	25 Jul.	Cap.	Yellow, very slow
Aquarids	15 Jul.-15 Aug.	28 Jul.	Aqr.	Slow, long paths
Pisces Australids	15 Jul.-20 Aug.	30 Jul.	PsA.	
Capricornids	15 Jul.-25 Aug.	1 Aug.	Cap.	Yellow
Aquarids	15 Jul.-25 Aug.	6 Aug.	Aqr.	
Perseids	25 Jul.-18 Aug.	12 Aug.	Per.	Fast, fragmenting
Cygnids	18-22 Aug.	20 Aug.	Cyg.	Bright, exploding
Orionids	16-26 Oct.	21 Oct.	Ori.	Fast, persistent
Taurids	10 Oct.-30 Nov.	1 Nov.	Tau.	Slow, brilliant
Leonids	15-19 Nov.	17 Nov.	Leo.	Fast, persistent
Phoenicids		4 Dec.	Phe.	
Geminids	7-15 Dec.	14 Dec.	Gem.	White
Ursids	17-24 Dec.	22 Dec.	UMI.	

Above: *principal annual meteor showers.* ***Right:*** *the Leonid meteor shower photographed in 1966. All the meteors appear to radiate from one point.*

sphere at a speed of anything up to several tens of miles per second, resulting in a sometimes brilliant streak of light seen against the background of stars. This effect is known as a meteor, or – incorrectly – as a shooting star. Some particles are large enough to partially survive entry through our atmosphere. These are meteorites. Very large meteorite falls are rare, yet when they occur they can form a crater at the place of impact. A notable example is the Arizona Meteorite Crater, a large depression over half a mile in diameter and around 650 feet (200 m) deep, which is the result of a meteorite fall that occurred many thousands of years ago. The study of meteorites has placed their age at around 4,500 million years.

Although meteors can appear at any time and from any direction (sporadic meteors), there are certain times of the year when meteor showers occur and a relatively large number of meteorites are seen to emerge, apparently from the same point in the sky. These showers have their origins in comets, which shed their material as they approach the Sun. Because the particles from these comets are all travelling in paths parallel to each other, they all seem to radiate from one particular point in the sky. This point is known as the radiant, and meteor showers are named after the region in the sky which contains the radiant.

Observing Comets

Although a large number of comets have recognized orbits, the majority of them are faint, and telescopes are required to see them. But one or two comets per year are generally visible with moderate optical aid, such as binoculars or a small telescope. Occasionally there may be a surprise visitor to our skies which is visible to the naked eye, such as IRAS-Araki-Alcock. This particular comet generated much public interest as it made an appearance in May 1983. Once you have plotted the position of a comet on a star chart, sweep the area of sky for a degree or two either side of the expected location, and remember that you are looking for a patch of light rather than a starlike point. Once found, note how the comet moves against the background of stars, this movement normally being detectable over a period of only a few hours.

Observing Meteors

The observation of meteors is one of the most straightforward and rewarding projects available to the amateur. The equipment required to carry out a meteor watch is basic; a sleeping bag and a sun lounger, a flask of coffee to keep you active, a writing pad and something to write with and an accurate watch. Meteor watches can be carried out on any clear (and preferably moonless) night of the year. A dark sky is preferred, and a countryside location well away from city lights will mean that fainter meteors will be visible.

If a shower watch is carried out there are a number of details that should be noted, including the starting and finishing times of the watch. To serve any useful purpose a meteor watch should be of at least an hour's duration, preferably more. If a group carries out a watch each member can watch a particular area of sky, and it is more likely that a greater number of meteors will be seen. During the watch, keep a count of the number of meteors seen and make a note whether they are shower or sporadic. If a meteor has any colour, make a note of it, and also record details of any trails that a meteor may leave behind itself, which will take the form of luminous glows along the meteor track.

Eclipses and Occultations

Lunar Eclipses

A lunar eclipse takes place when the Moon passes into the Earth's shadow. As with all objects in the Solar System the Moon is illuminated by the light from the Sun. However, during an eclipse this light is temporarily cut off and the lunar surface is plunged into darkness. It is only on very rare occasions, however, that the Moon becomes completely invisible, as a small amount of sunlight is normally refracted onto the lunar surface by the atmosphere surrounding our planet.

The shadow that is thrown onto the lunar surface has two quite distinct regions. The dark central region is known as the umbra, and to an observer standing on the lunar surface in this area the Sun would be completely hidden from view. The lighter area of partial shadow surrounding the umbra is the penumbra, from which area the Sun would only be partly hidden. These different areas of shadow give rise to different types of eclipse. A total lunar eclipse occurs when the Moon passes through the umbra of the Earth's shadow; a partial eclipse occurs if only a part of the Moon enters this region. When the Moon misses the umbra altogether and simply passes through the penumbra, a penumbral eclipse takes place. These are often difficult to detect as the effect of darkening on the lunar surface is only very slight.

The ancient Greeks were well aware of how lunar eclipses happened, but it was during a particular eclipse in the 4th century BC that Aristotle observed that the shadow of the Earth on the lunar surface was curved, and concluded that the Earth must be spherical.

Solar Eclipses

Although the Sun is considerably larger than the Moon it is also much further away, so that the apparent diameter of each is roughly the same. A solar eclipse takes place when the Moon passes between the Sun and Earth, and they can be total, partial or annular. A total solar eclipse occurs when the lining up of the three bodies is exact and the solar disc is completely obscured by the Moon; a partial eclipse takes place when only a fraction of the Sun is hidden, Annular eclipses also take place when the lining up is exact, but they happen when the Moon is at its furthest point on its orbit, when its apparent diameter is somewhat smaller than that of the Sun. The result is that the Sun will be visible as a bright ring – or *annulus* – around the Moon.

Because of the combined effects of the Earth's axial rotation and the movement of the Moon through its orbit, during a total eclipse the Moon's umbra sweeps across the Earth's surface along what is known as the "path of totality". Anybody observing from within this area will see a total eclipse, while those on either side will witness only a partial eclipse. It may be that the lining up of Sun, Moon and Earth is not exact, and on these occasions a partial eclipse may take place with no associated totality.

During any year, the maximum number of solar eclipses is five, while that of lunar eclipses is three. Solar eclipses are only visible from regions of the Earth on or near the path of totality, whereas lunar eclipses can be seen from any location of the hemisphere facing the Moon at the time. The occurrence of eclipses is subject to a definite cycle lasting 18 years and 11 days, which is known as the Saros period.

Occultations

An occultation takes place when one

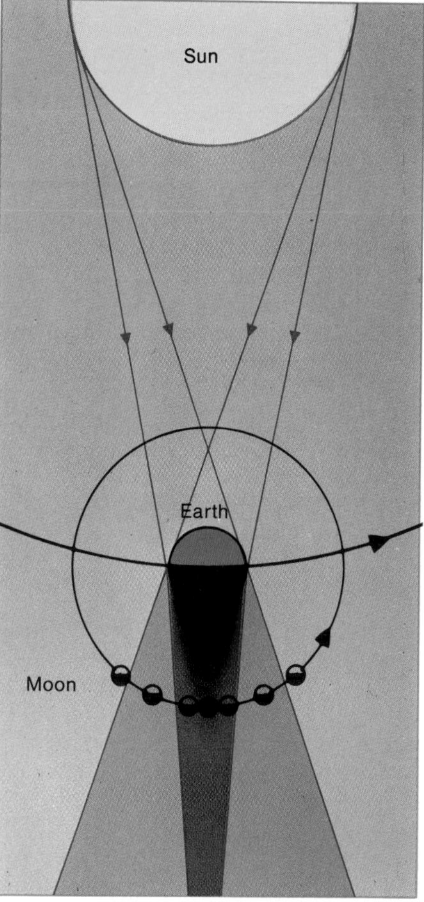

Top: *the lunar eclipse of June 1964. The Earth's shadow is shown crossing the lunar disc in the first four pictures. The fifth shows totality and the last was taken shortly after totality ended.*
Above: *theory of a lunar eclipse when the Moon enters the Earth's shadow.*

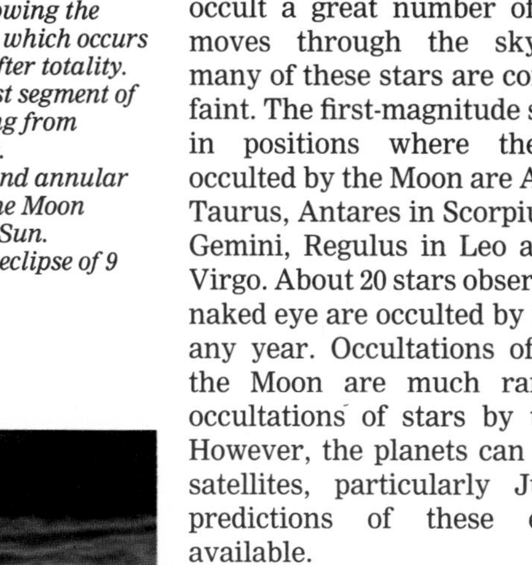

***Above:** total solar eclipse of 26 February 1979, with the Sun's corona visible around the lunar disc.*
***Above right:** total solar eclipse of 16 February 1980, showing the Diamond Ring effect, which occurs just before and just after totality. In this picture the first segment of the Sun is reappearing from behind the lunar disc.*
***Left:** theory of total and annular solar eclipses when the Moon passes in front of the Sun.*
***Below:** partial solar eclipse of 9 April 1986.*

celestial body, such as the Moon, passes in front of another, for example, a star or planet. The Moon has quite a large apparent diameter and can occult a great number of stars as it moves through the sky, although many of these stars are comparatively faint. The first-magnitude stars that lie in positions where they can be occulted by the Moon are Aldebaran in Taurus, Antares in Scorpius, Pollux in Gemini, Regulus in Leo and Spica in Virgo. About 20 stars observable by the naked eye are occulted by the Moon in any year. Occultations of planets by the Moon are much rarer, as are occultations of stars by the planets. However, the planets can occult their satellites, particularly Jupiter, and predictions of these events are available.

The observation of occultations is important in determining the precise orbital motions of the Moon and planets. Although the positions of stars are known with great accuracy, the exact orbit of the Moon is not. But, because the Moon has no atmosphere, there is no perceptible dimming of the light from the star prior to occulation, and as it disappears behind the lunar disc it seems to "snap out" suddenly. Therefore, by precisely timing these events, it is possible to calculate the exact location of the Moon at the time.

Once you have witnessed an occultation you will agree that there is something quietly mysterious about watching the light from a distant star suddenly vanish behind the Moon.

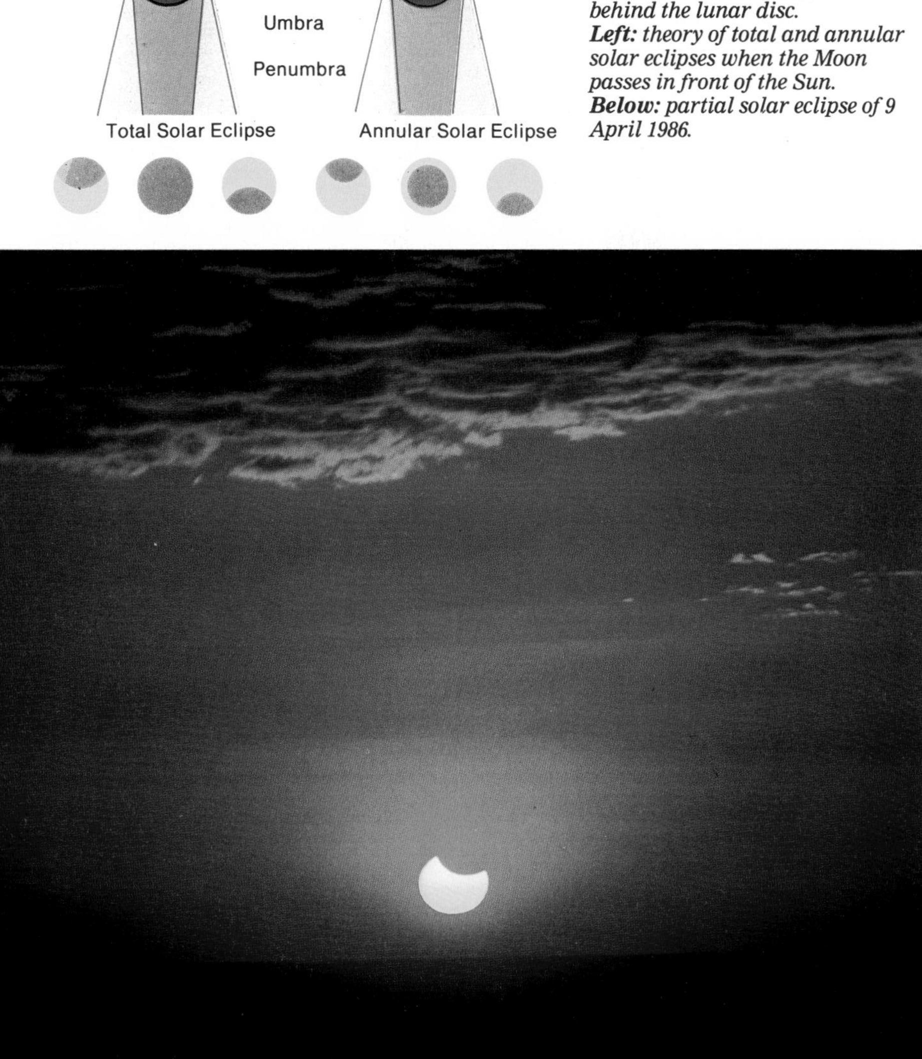

Heavenly Lights

Aurorae

The Sun emits a constant stream of charged particles, known as the solar wind. Travelling at speeds of around 375 miles (600 km) per second they eventually reach the Earth's environment and enter the Van Allen Belts – regions above the Earth in which protons and electrons are in constant movement between the two magnetic poles. An increase in the density of the solar wind, such as that which occurs during times of maximum sunspot activity, can overload the Van Allen Belts. Charged particles are then ejected into the upper atmosphere, where they react with air particles, causing them to give off radiation which we see as the aurorae.

Auroral activity is concentrated by the Earth's magnetic field into two areas known as the northern and southern auroral zones, which are centred on the Earth's magnetic poles. The best displays of aurorae are seen from Greenland, Iceland and the northernmost point of the North American continent, and in the south the finest shows take place over locations well removed from populated land masses.

Occurring at heights of between 65 miles (100 km) and 650 miles (1,000 km) above the Earth's surface, aurorae can be any of a number of different colours, including red, green and blue. They can also take on any of a number of different forms. Arcs, crowns, bands and rays can appear in the sky when solar activity is particularly strong, and it is at these times that those in more populated areas can catch sight of the aurorae.

The name "aurora borealis", or northern dawn, describes fairly accurately the appearance of aurorae from places some distance from the auroral

***Left:** in the event of an increase in solar activity the charged particles from the Sun (**A**) entering the Van Allen Belts (**B**) spill into the upper atmosphere (**C**) and give rise to aurorae.*
***Below:** an auroral display photographed from Lincolnshire.*
***Right:** an auroral display photographed from Cumbria.*
***Lower right:** the Zodiacal Light is produced by the reflection of sunlight from tiny particles scattered along the plane of the Solar System.*

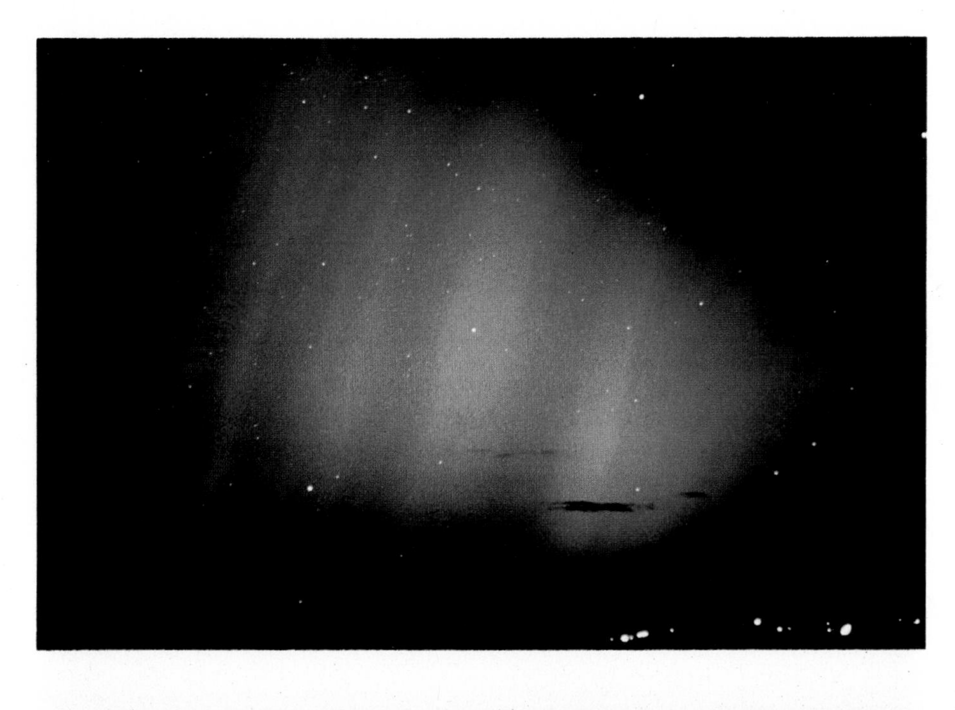

zones, where a glow on the poleward horizon is the most common sight. A dark sky will increase the chances of seeing an auroral display, but some of the really bright displays can even be observed from city areas, though this is not common. The relatively unimpressive horizon glow of a distant auroral display can extend itself upwards to form an arc, stretching from east to west. Vertical ray formations can be produced to form a rayed arc. These rays can then stretch themselves away from the arc, forming a rayed band, resembling a giant curtain waving in the wind. These ray formations follow the lines of force from the Earth's magnetic field and sometimes appear to collect together at the zenith, producing a coronal aurora. Patches of auroral light sometimes collect in otherwise clear skies, and vertical rays, resembling enormous searchlight beams, can reach up from the horizon. Aurorae can also form bands, rather like arcs in appearance but with twists along their length giving the effect of a ribbon in the sky.

Observation is best with the naked eye. Telescopes are of no use and binoculars will serve only to locate stars to help fix the positions of auroral displays. Another advantage is that your hands are free to record observations. Various details should be noted, including the shape and form of the display, the date and time it took place, observing conditions and the presence of either clouds or moonlight. Finally, mention can be made of brightness. This can be given as a relative value, the display being compared with the Milky Way, moonlit cirrus clouds, cumulus clouds or even brighter. There are many who have never seen an aurora, and if you are lucky enough to see a display enjoy it – it may be the one and only chance you will get.

Zodiacal Light

The Zodiacal Light is visible as a faint cone of light reaching up from the horizon either in the east before sunrise or in the west after sunset, and is best seen in either spring or autumn, when the ecliptic is closest to being perpendicular to the horizon. It is easier to see from sites on or near the equator, where the ecliptic is virtually at right angles to the horizon, although it can be observed from elsewhere under the right conditions. To observe it you must be well away from any artificial lights and the sky must be clear and moonless. The only way it will be seen is with the naked eye. The Zodiacal Light is at its brightest when the Sun is around 20° below the horizon, which corresponds to an interval of two hours during spring and autumn. The Zodiacal Light may be as much as 20° wide at its base and may reach to a height of between 40° and 60° above the horizon. If seen, its outline and position in relation to background stars may be noted. It is, however, very difficult – though not impossible – to capture on film. If photography is attempted, use a very fast film, a wide-angle lens, a camera mounted on a tripod and exposures of 30 seconds to several minutes. The observing site must be dark and completely free of light pollution.

The Gegenschein

The Gegenschein is a faint elliptical patch of light situated at a point in the sky exactly opposite to the Sun. Exceedingly faint, it has been described as the most elusive object in the sky. Its dimensions are in the region of 10° wide and 20° long and it is best seen with the naked eye when the anti-solar point is furthest from the region of the Milky Way, during the periods February to April and September to November.

The Zodiacal Band

This is an extremely faint, parallel-sided band of light running along the ecliptic. Its width is normally between 5° and 10°, but it is rarely seen.

The Stars

The Birth of Stars

Stars are formed inside clouds of gas and dust, by a process known as fragmentation – the first step towards the formation of a star cluster. As the fragments collapse, their densities and temperatures become greater until thermal pressure builds up to such an extent that it prevents any further collapse, and a number of stable regions known as protostars result.

The next stage of development will depend on the mass of the protostar. If this mass is comparable to that of our Sun, a hot, central region will form. The outer layers will continue to be drawn inwards as the temperature at the core gradually increases to the point when the temperature reaches a value of around 10 million degrees K, when nuclear reactions are triggered off and the hydrogen in the core is converted into helium, preventing any further gravitational collapse and allowing the star to become stable and to produce light and heat for about 10,000 million years before further changes take place. However, stars with masses greater than the Sun consume their hydrogen fuel at a much quicker rate and have much shorter lifetimes.

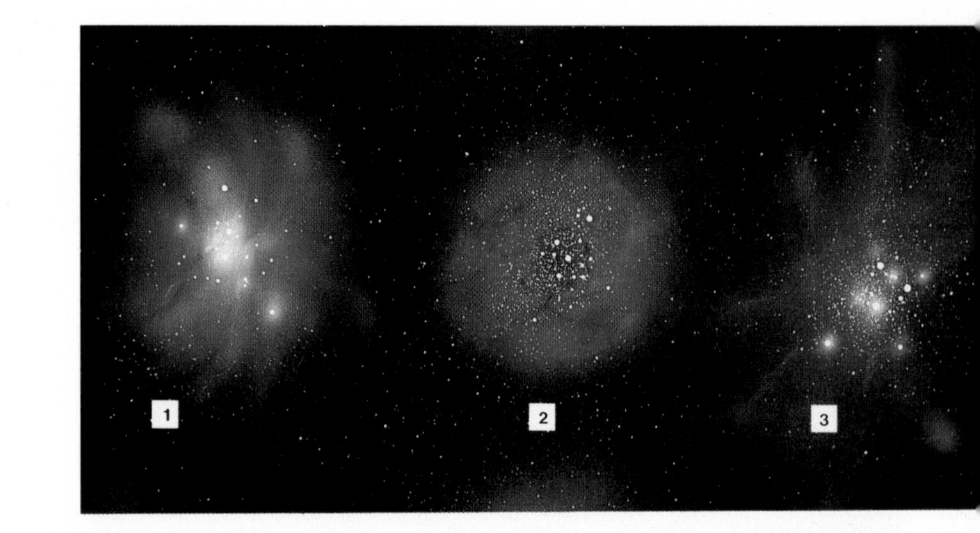

Red Giants

The Sun will reach the stage when the core hydrogen supply is depleted and the helium-producing reactions come to a virtual halt some 5,000 million years from now. Its helium-rich core will now be surrounded by a zone in which hydrogen is still being converted into helium, but this zone will gradually expand until it reaches a point where the temperature is too low for the reactions to continue. While this is happening the core will contract through the action of gravity, which in turn will cause its internal temperature and pressure to increase. Once this temperature reaches a value of 100 million degrees K a further round of nuclear reactions is triggered and the helium is converted into carbon. The radiation pressure from the rejuvenated core, coupled with that of the helium-producing shell, brings about a major change. The outer layers expand to many times their original size and the star becomes a red giant.

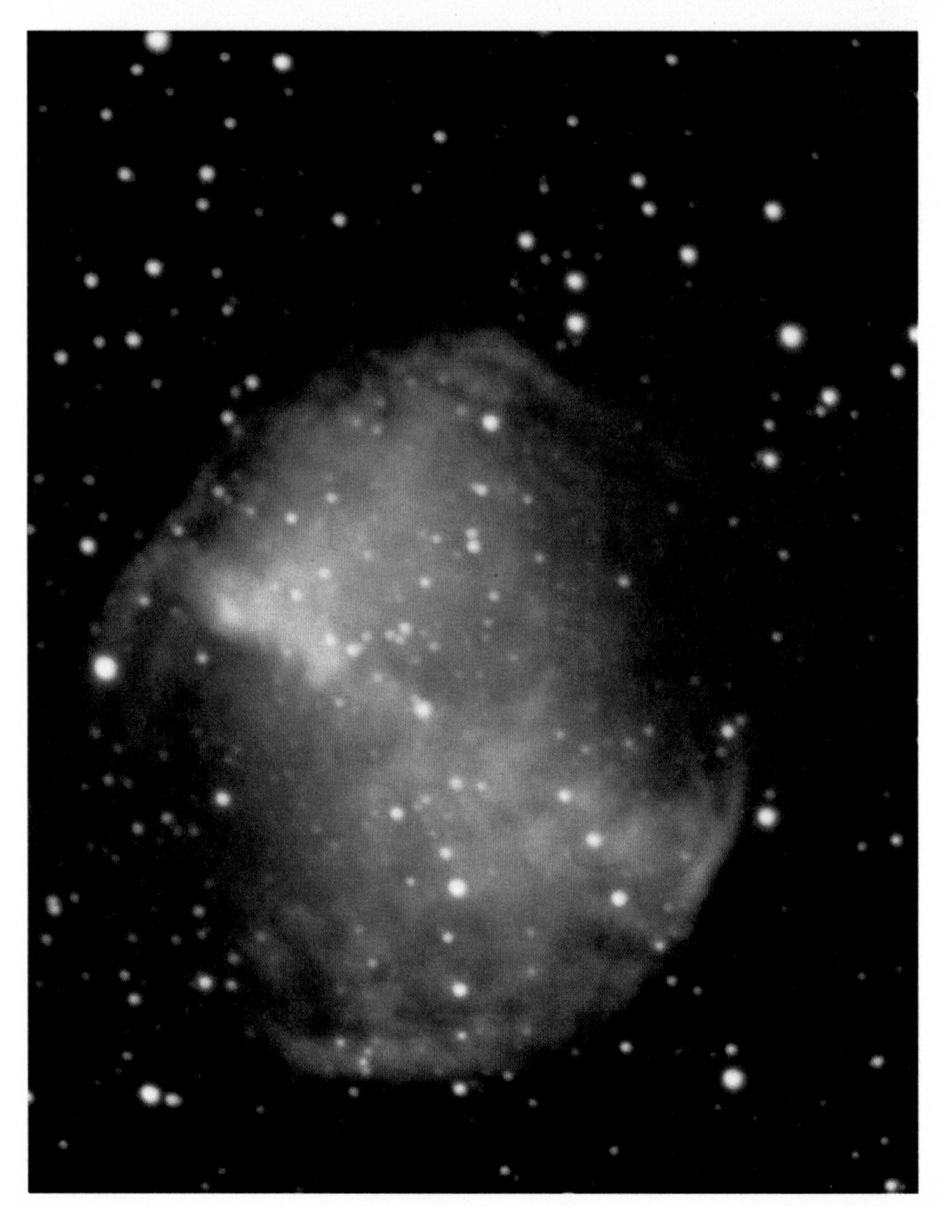

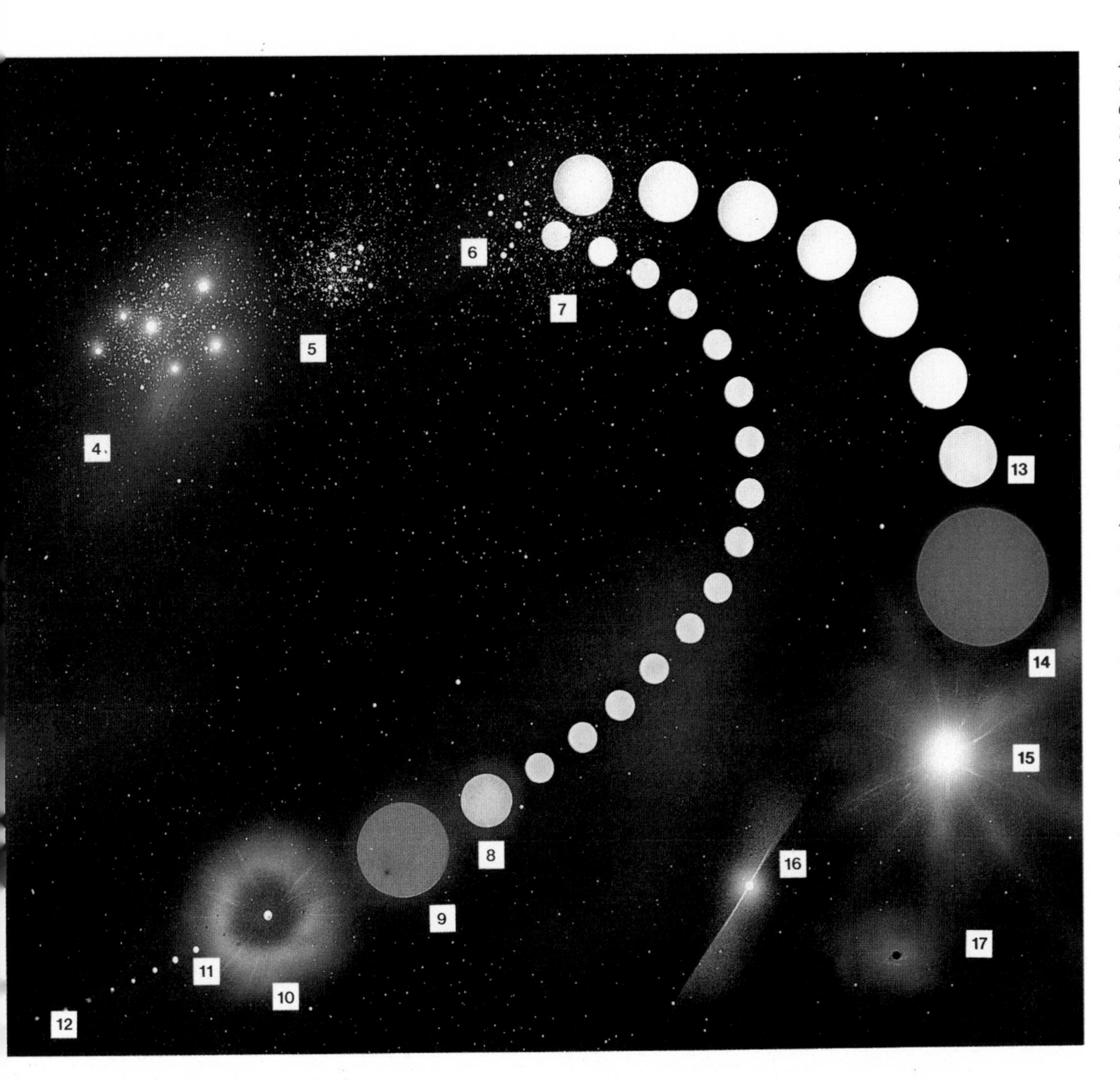

Left: *star formation begins inside nebulae* ***(1)*** *which collapse, producing increased temperatures* ***(2,3)****. Stars begin to shine, their energy blowing away any remaining gas* ***(4)*** *leaving a star cluster* ***(5)****. Perturbations both within the cluster and from non-cluster stars cause the cluster to eventually become a loose stellar association* ***(6)****. A solar-type star* ***(7)*** *evolves differently to its more massive counterparts. Entering the Main Sequence it remains on it until the hydrogen supply diminishes and the star expands* ***(8)*** *to become a red giant* ***(9)****. Its outer layers are ejected to form a planetary nebula* ***(10)****, while the remnant collapses to form a white dwarf* ***(11)****. It continues to shine feebly, eventually cooling to become a cold, dark black dwarf* ***(12)****. More massive stars have a shorter period on the Main Sequence, after which* ***(13)*** *they become supergiants* ***(14)*** *and may undergo a supernova explosion* ***(15)****. The mass of the star dictates whether it ends its life as a neutron star or pulsar* ***(16)*** *or produces a black hole* ***(17)****.*
Far left: *the Dumbbell planetary nebula in Vulpecula.*

Planetary Nebulae

Planetary nebulae are not true nebulae in the strict sense of the word. Whereas the word "nebula" is taken to describe a cloud of gas and dust in space, planetary nebulae are indicative of a particular stage in the evolution of certain stars. A planetary nebula is formed as a red giant undergoes its final phase of nuclear reactions and the star's outer layers are blown away. Nearly a thousand examples are known, the most famous being the Ring Nebula in Lyra.

White Dwarfs

A star may remain a giant for millions of years, but eventually all the nuclear fuel will be used up and no further reactions will take place, and the star will collapse under its own gravity. Although all nuclear reactions will eventually have ceased, the gravitational energy produced during the collapse of a star of less than 1.4 solar masses – known as the Chandrasekhar limit – would be converted into heat and the white dwarf would continue to shine, albeit rather feebly. However, all the remaining energy would eventually radiate into surrounding space and the star would become a black dwarf.

Stars above the Chandrasekhar limit may also collapse into white dwarfs provided that they are able to shed some of their mass beforehand. The most obvious way for this to happen would be for the outer layers of the star to be ejected into space, thereby creating a planetary nebula. Once the nebula has formed the star itself will start the collapse which will result in it becoming a white dwarf.

Neutron Stars and Pulsars

Stars that lie above the Chandrasekhar limit but below three solar masses undergo a gravitational collapse that takes them past the white dwarf stage. This collapse is so powerful that protons and electrons inside the star are smashed together to form neutrons, so forming a neutron star. The material making up the neutron star is so incredibly dense that a single cubic inch would have a mass of several million tons. If the Sun could be compressed to a similar density its diameter would be reduced to only 18 miles (30 km).

Until the 1960s neutron stars existed only in theory. As far as astronomers were concerned neutron stars represented the hypothetical end of stars above the Chandrasekhar limit, although there was no observational evidence available. However, all this was about to change. In 1967 radio astronomers picked up bursts of radio emission from an area of sky in which there was no visible source. The signals were being transmitted on an extremely regular basis, so regular in fact that their frequency could be timed to 1/10,000,000 of a second, the actual period being 1.3370109 seconds. It was thought that these bursts might originate from an object such as a star which was pulsating. The new object was christened "pulsar", and a search through previous records revealed three more. Eventually many more were discovered, notably one at the heart of the Crab Nebula in Taurus.

In 1054 the appearance of a new and extremely bright star in the constellation of Taurus was recorded by Chinese astronomers. What they actually saw was a supernova. Supernovae take place as the result of the gravitational collapse of a very massive star. This collapse is so forceful that a violent nuclear reaction takes place which causes the star to explode, throwing its outer layers into surrounding space. For a time it can become over a billion times as bright as the Sun. After the star explodes, its core collapses into a neutron star. The Crab Nebula is actually the scattered remnants of the 1054 supernova. The pulsar in the centre of the Crab Nebula has now been identified as a rapidly spinning neutron star which emits radio pulses synchronous with its rotation at the incredible rate of 30 per second! Over 300 pulsars have now been catalogued, although only three have been identified optically. One is the Crab pulsar, the second is that associated with the Vela supernova remnant and the third is PSR 1937 + 214 in the constellation of Vulpecula. In each case the star has been found to flash on and off in time with the radio pulses.

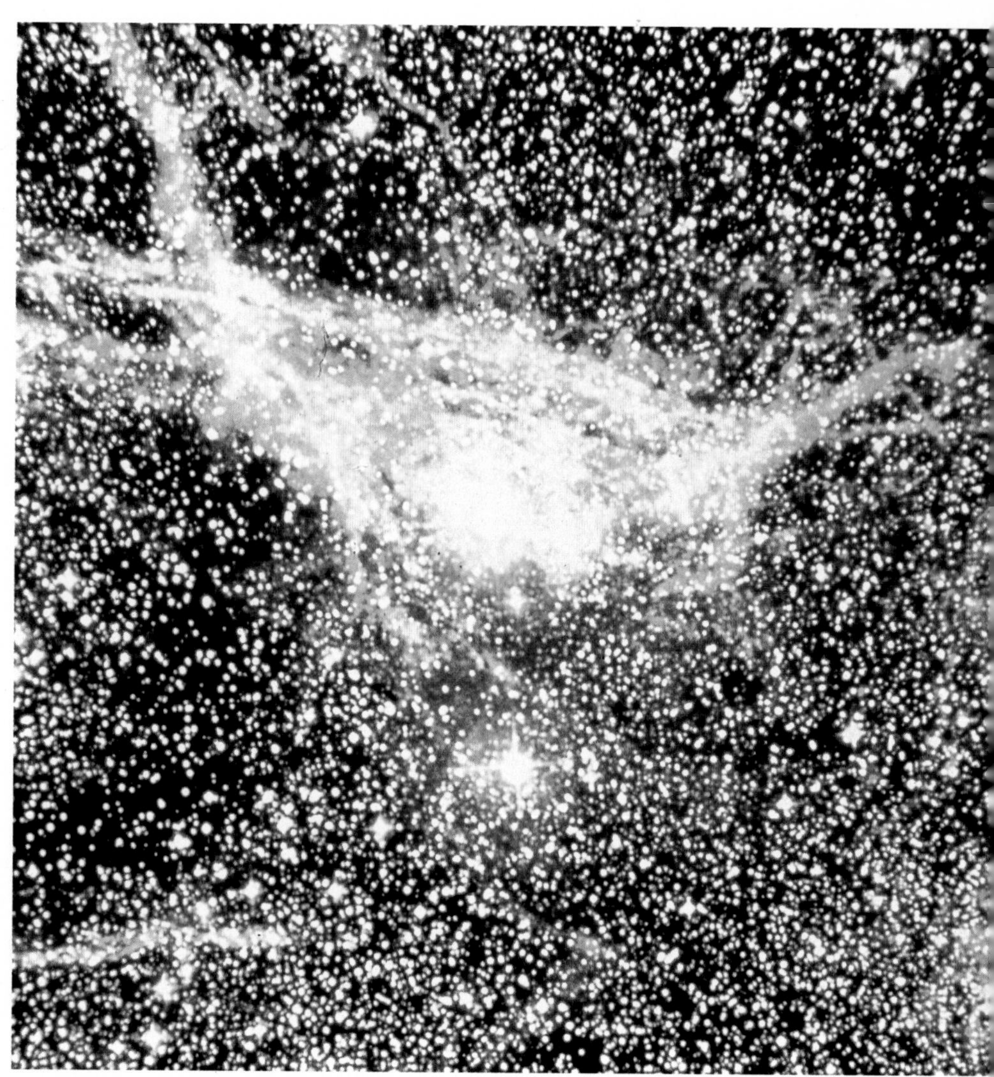

Black Holes

Stars with a pre-supernova mass of eight solar masses or more may collapse beyond both the white dwarf and neutron star stages. The star shrinks under an incredible gravitational pull, which crushes it into a sphere of increasing density and decreasing size. As the density builds up the escape velocity at the surface of the star gets higher and higher until it eventually exceeds the speed of light. As this point is passed light rays, which up until now have been able to escape from the star, are bent back on themselves and are unable to leave. A notional nearby interstellar explorer would be able to watch the star disappear from sight. The collapse continues until it can go no further. Although the gravitational field surrounding the collapsed star will not permit light to escape, the pull of gravity weakens with increasing distance. Eventually there will be a point at which light can break free. This is the "event horizon", enclosing a zone forever hidden from our view. This zone is known as a black hole.

By definition we are unable to see black holes. Yet the study of binary stars – pairs of stars in orbit around each other – may provide us with some clues. Close examination of binary stars enables us to calculate their masses. Those systems which contain either a white dwarf or neutron star can emit bursts of X-ray radiation, when material is dragged from the surface of the largest star by its denser companion. This gas forms a ring, known as an "accretion disc", around the more dense component. The material in the accretion disc gets hotter and hotter and eventually emits X-rays. The vast majority of these systems involve neutron stars, although it is now thought that a number of them play host to black holes. In 1971, an extremely bright X-ray source was discovered in the constellation of Cygnus. When astronomers examined the area at optical wavelengths they found a hot, blue supergiant star which appeared to be in orbit around an invisible companion. This companion "star" was shown to have a mass around five times that of the Sun. The system was christened Cygnus X-1 and it is thought that the unseen component is actually a black hole which throws out X-rays as material from the supergiant star is pulled down through the event horizon. Other, similar systems have been detected and yet new observing techniques may be required before we can even hope to unlock the secrets of the mysterious phenomena known as black holes.

Stellar Distances

It would be meaningless to try to express the vast distances of stars in miles, so astronomers use another unit – the light year. This is equal to the distance that light travels in a year, which comes to a value of around 6 million, million miles. On this scale the Sun is just over 8 light minutes away, Pluto around 5½ light hours and the nearest star, Proxima Centauri, 4.3 light years.

Stellar distances are measured by different means, the closest stars having their distances measured by trigonometrical parallax, which makes use of the apparent angular shift of the star against the background of more distant stars when viewed from two different positions. These measurements are taken at intervals of six months, when the Earth will be at opposite points in its orbit around the Sun. The diameter of the Earth's orbit is used as a base-line and, using the value of shift, it is fairly easy to assess the distance to the star using simple trigonometry. In 1838, Friedrich Bessel used this method for the first time when he measured the distance of the star 61 Cygni, which he estimated as lying at a distance of around 11 light years.

However, the more distant a star is, the smaller will be its angle of parallax, and this angle becomes difficult to measure for stars more than 70 light years or so away. In these cases a comparison is made between the distant star and one of a similar type that lies at a known distance. Assuming that stars of similar types will have identical brightnesses, astronomers can estimate how far away the remote star is by comparing its actual and apparent brightnesses. The comparison method relies on astronomers being able to identify both the type and brightness of distant stars.

Magnitudes

Even a casual glance at the night sky will show that some stars are very bright while others hover on the limit of naked-eye visibility. As long ago as 150 BC, the Greek astronomer Hipparchus devised a system whereby the stars were classed according to their brightness, with the brightest stars

Left: *the Vela supernova remnant was created in a supernova explosion which took place around 10,000 years ago.*

Left: *a black hole represents the theoretical final stage in the evolution of a very massive star. Although black holes cannot be detected optically, those that are members of binary systems reveal themselves through violent X-ray emissions thrown out as material from the companion star is drawn into its gravitational trap.*

Above: *possible model of the Cygnus X-1 system where material from a supergiant star* ***(1)*** *is drawn away* ***(2)*** *to form an accretion disc around its black hole companion* ***(3)****. This material is heated and emits X-rays before being pulled in by the intense gravity of the superdense collapsed star at the heart of the black hole.*

being given a magnitude value of 1 and the faintest a value of 6. This system is still used today, although with certain refinements. Magnitudes can be estimated to within 0.01 with modern instruments. Also, the brightest stars have had to be designated zero or even negative values to fit in with the current scale. However, it isn't just stars that have had their magnitudes assessed. For example, the Sun has a magnitude of minus 26.7, while that of the full Moon is minus 12.7. Sirius, the brightest star in the sky, has a magnitude of minus 1.4, although Venus, the brightest of the planets, can reach a magnitude of minus 4.4 when suitably placed. Pluto, on the other hand, hovers at around 15th magnitude, although this can vary according to its distance from the Sun. Uranus can just become visible to the naked eye at magnitude 5.6. A pair of 7 × 50 binoculars will show stars down to 8th or 9th magnitude, although in comparison the world's largest telescopes will reveal objects such as distant galaxies down to about 25th magnitude. A difference of 1 magnitude between two stars is equivalent to a difference in brightness of 2.512. For example, a star of magnitude 3 is 2.512 times as bright as one of magnitude 4, and (2.512 × 2.512) = 6.31 times as bright as a star of magnitude 5 and so on. In 1856, Pogson standardized the magnitude system and he estimated that a star of 1st magnitude was 100 times brighter than one of 6th magnitude. It follows on from this that the difference between successive magnitudes was $\sqrt[5]{100}$, or 2.512.

The above system classifies celestial objects according to their *apparent magnitude*, which is a measurement of their brightness in visible light only. However, stars emit radiation at many different wavelengths and these are taken into account when the *bolometric magnitude* is assessed. It is sometimes difficult to calculate the bolometric magnitude of a star due to the fact that the Earth's atmosphere absorbs certain types of radiation. There is also *absolute magnitude*, which is the magnitude a star would have at a distance of 10 parsecs, or 32.6 light years. A parsec (3.26 light years) is the distance at which a star would have an angle of parallax of 1 second of arc. At this distance the Sun would shine with a magnitude of 4.83, while Sirius would be much brighter at magnitude 1.3. Meanwhile, Polaris, the Pole Star, is situated at a distance of several hundred light years and has an apparent magnitude of 1.99, although its absolute magnitude is minus 4.6. It can be seen that the absolute magnitude of a star is a far better indication of its actual luminosity.

Colours of Stars

Stars can be seen to have many different colours, a prominent example being the bright orange-red Arcturus in the constellation of Boötes, which contrasts sharply with the nearby brilliant white Spica in Virgo. Our own Sun is yellow, as is Capella in Auriga. Procyon, the brightest star in Canis Minor, also has a yellowish tint. To the west of Canis Minor is the constellation of Orion the Hunter, which boasts two of the most conspicuous stars in the whole sky; the bright red Betelgeuse and Rigel, the brilliant blue-white star that marks the Hunter's foot. Other stars with conspicuous colours include orange-red Aldebaran in Taurus and Mu Cephei in the constellation of Cepheus. Mu Cephei, or the Garnet Star, is probably the reddest star visible to the naked eye in the

the northern skies, and binoculars will bring out the colour very well.

The colour of a star is a good guide to its temperature. The hottest stars are blue and blue-white, which have surface temperatures of 20,000 degrees K or more. The Sun is a fairly average star with a temperature of around 6,000 degrees K, while red stars are much cooler still, their surface temperatures being in the region of only a few thousand degrees K. Betelgeuse in Orion and Antares in Scorpius are both red giant stars that fall into this category.

Stellar Sizes

Stars can also differ greatly in size. The Sun is a fairly average star with a diameter of 870,000 miles (1,400,000 km), although this is tiny in comparison to supergiant stars such as Antares, which is some 280,000,000 miles (450,000,000 km) across. The smallest stars known include white dwarfs and neutron stars. Examples of white dwarfs include the dim companions of Sirius and Procyon, which have diameters of only a few thousand miles. Neutron stars are even smaller and are typically only a few miles across.

Spectral Classes of Stars

If the light from a star is passed through a prism it will be split up into its constituent colours, ranging from long-wavelength red through orange, yellow, green and blue to short-wavelength violet. This rainbow effect is known as a spectrum and if the source of light is a high-density gas, such as that found within a star, a *continuous spectrum* will be produced with an unbroken sequence of colours. However, low-density gas, typically that which surrounds the surface of a star, will produce an *emission spectrum*, which takes the form of a series of individual bright lines, each line being produced by the effect of a particular element. When a star is examined spectroscopically, both types of spectrum are produced with the emission spectrum superimposed on the continuous spectrum. However, the lines in the emission spectrum are seen as dark bands in contrast to the brighter spectrum underneath, and indicate the presence of particular elements within the atmosphere of the star. So, although the stars are remote, astronomers can tell a great deal about their compositions by studying their spectra.

There are different spectral classes of star, each one of which is classified by a letter from the sequence O, B, A, F, G, K, M, R, N and S. The spectral class of a star is a clear indication of its temperature, with the hottest stars being of spectral type O. The cooler stars are classed as M, R, N or S. Each spectral class can be further divided by a number, ranging between 0 and 9, after the letter. The lower the number, the higher the temperature of the star. For example, the Sun is classed as a G2-type star, and is hotter than a star of type G5 and so on. Since the middle of the nineteenth century it was known that stars could be split into different spectral types, and in 1863 the Italian astronomer Angelo Secchi published the first classification of spectra,

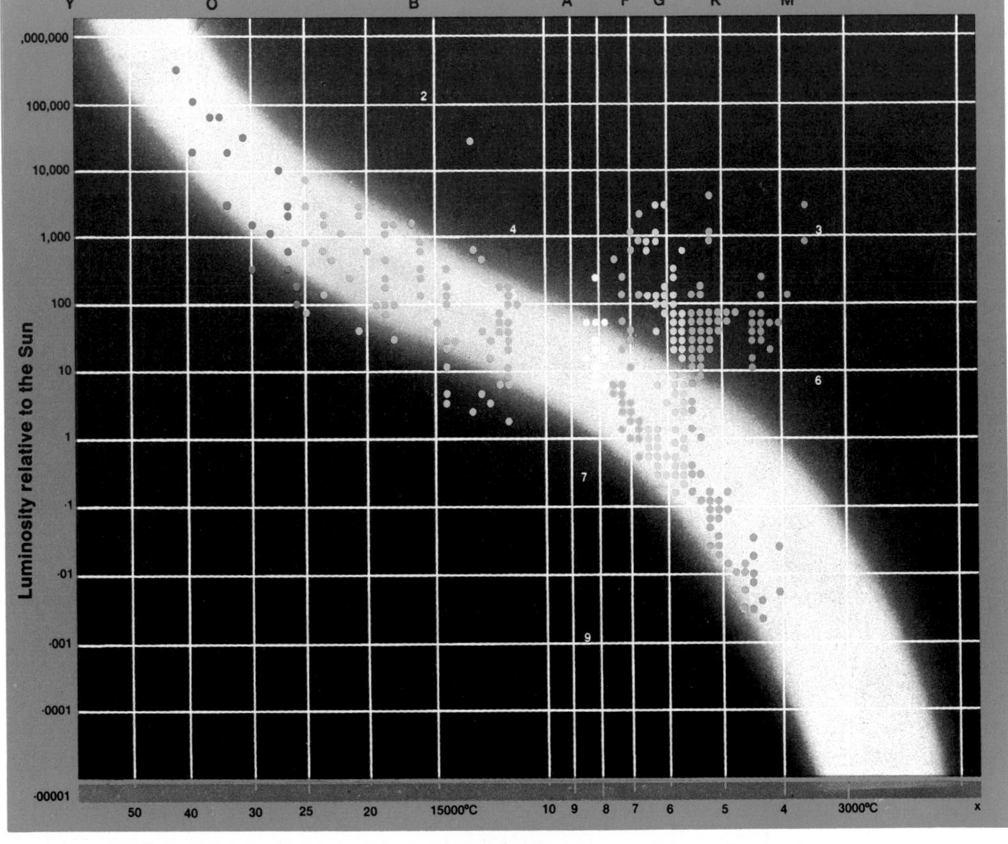

***Far left:** the conspicuous red Betelgeuse (top) and pure white Rigel (bottom) stand out well in this view of Orion. **Left:** the Hertzsprung-Russell diagram plots stars according to their surface temperatures and spectral class (horizontal axis, x) and luminosities relative to the Sun (vertical axis, y). Even a casual glance will show that many stars are concentrated into a band ranging from very hot and luminous stars at upper left **(1)** through to the relatively cool, dim red dwarfs at lower right **(8)**. This is the Main Sequence, and it contains around 90 per cent of all known stars, including our own Sun. Other star types shown are the supergiants and giants **(2,3)**, Cepheid and RR Lyrae variables **(4,5)**, sub-giants **(6)**, sub-dwarfs and white dwarfs **(7,9)**. The H-R diagram is the result of work carried out earlier this century by the Danish astronomer Ejnar Hertzsprung and the American astronomer Henry Norris Russell, who came to similar but quite independently achieved conclusions regarding stellar constitution and evolution.*

which included a total of four classes. This was superseded in 1901 by Antonia Maury and Annie Jump Cannon of the Harvard College Observatory. They greatly expanded the list of spectral types and introduced the seven classes O, B, A, F, G, K and M. The addition of the three remaining classes – R, N and S – was made later. The system of classifying stellar spectra by these letters is known as the Harvard classification.

Double and Multiple Stars

Most of the stars that we see in the night sky are not alone in space but are actually members of multiple star systems, and a great number of these make up double or binary stars. From our vantage point on Earth we see many pairs of stars which only appear to be close to each other because they happen to lie in the same line of sight. These are known as *optical doubles* and are unlike the *binary stars,* which are comprised of two stars in orbit around their common centre of gravity. There are also *physical doubles,* which are made up of two stars that are bound together through gravitational attraction although no orbital motion has been detected.

Although a great number of binary systems have components that are clearly resolvable, there are some systems that have stars so close to each other that even the world's largest telescopes are unable to split the pair. These very close pairs are known as *spectroscopic binaries* and it is only through the use of the spectroscope that we are able to detect each individual star.

The spectrum of a star is made up of a continuous spectrum, superimposed upon which is an emission spectrum. If the star is moving away the wavelengths of light reaching us are stretched a little, and the dark lines of the emission spectrum are shifted towards the red or long-wavelength end of the spectrum. The situation is reversed for a star approaching us, in which case the wavelengths are compressed and the lines are shifted towards the blue or short-wavelength end. This process is known as "red shift" (or "blue shift") and is a good indicator of the velocity of a stellar

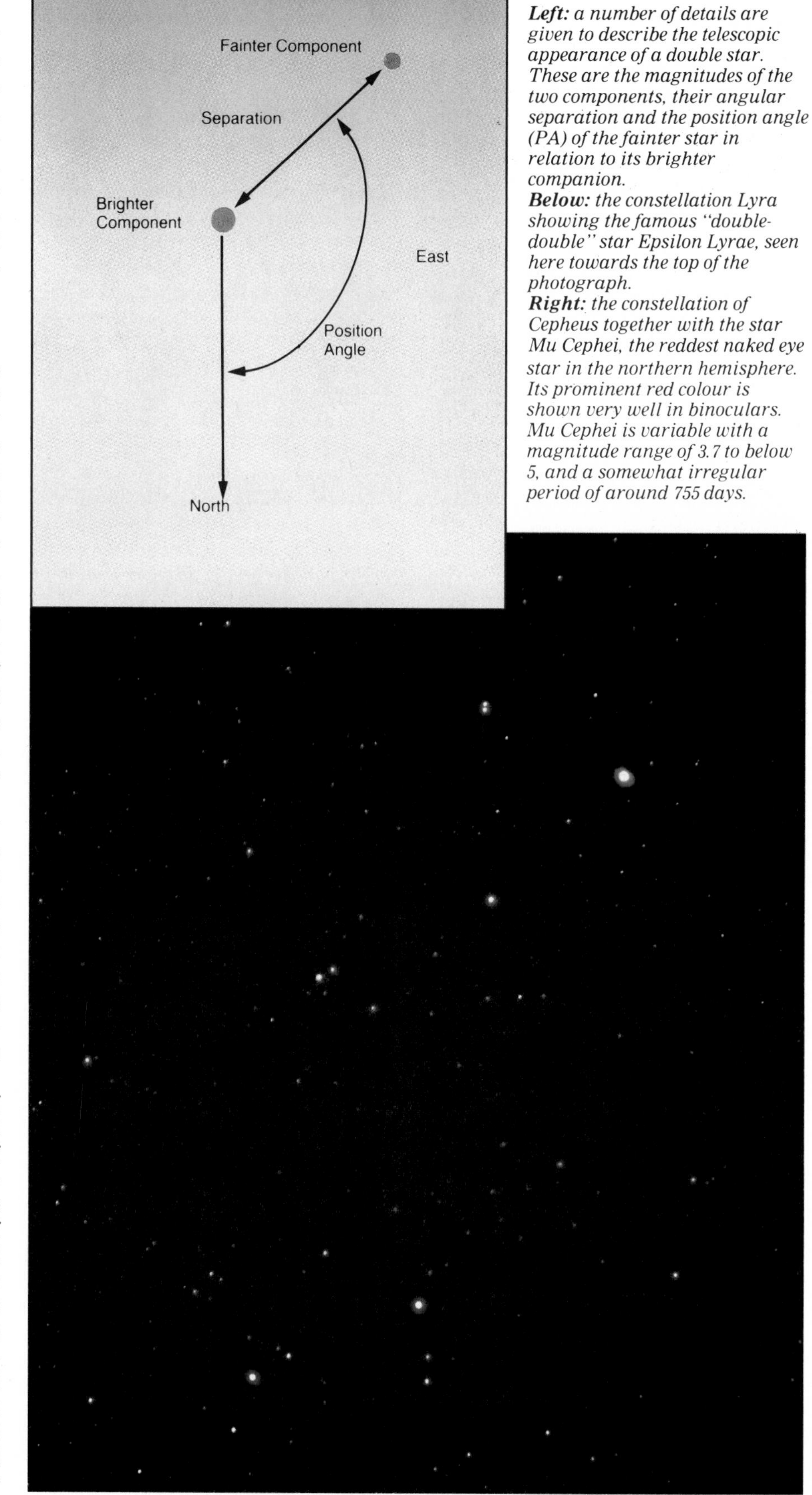

Left: *a number of details are given to describe the telescopic appearance of a double star. These are the magnitudes of the two components, their angular separation and the position angle (PA) of the fainter star in relation to its brighter companion.*
Below: *the constellation Lyra showing the famous "double-double" star Epsilon Lyrae, seen here towards the top of the photograph.*
Right: *the constellation of Cepheus together with the star Mu Cephei, the reddest naked eye star in the northern hemisphere. Its prominent red colour is shown very well in binoculars. Mu Cephei is variable with a magnitude range of 3.7 to below 5, and a somewhat irregular period of around 755 days.*

object either towards or away from us. The greater the velocity, the larger the shift of spectral lines. A similar effect takes place with sound waves. If a source of sound, such as a speeding train, is approaching us, the sound waves are compressed and the train has a higher pitch than when it passes and starts to move away. It is then that the pitch of the note drops due to the fact that the wavelengths of sound are stretched.

The idea that wavelengths of light would behave in this way was first suggested by the German mathematician Christian Doppler in 1842 and was expanded upon some years later by the French physicist Hippolyte Fizeau, who stated that the emission lines in the spectrum of a star would be seen to move in relation to the changes in wavelength.

The distance between the two components of a double star is expressed in seconds of arc (*see Our Changing Perspective*), regardless of whether the double is an optical, binary or physical system. The position of the stars relative to each other as seen from Earth is given as the Position Angle (PA). The fainter component is plotted against its brighter companion, the orientation being expressed in degrees from a position north of the brighter star through east, south and west. For example, if the fainter component is due south of its companion the PA of the system is 180°, if due west it is 270° and so on.

In a binary system, the point at which the two components are closest to each other is known as the periastron, while the furthest point is the apastron. These terms refer to actual orbital positions and not just their positions as viewed from Earth. In the case of a binary system, if the orbital motion of the fainter star is resulting in the PA increasing, the orbit is said to be direct. A retrograde motion is one that produces a decrease in the PA.

One of the most famous multiple star systems is Epsilon Lyrae, found roughly $1\frac{1}{2}°$ to the north-east of Vega in the constellation of Lyra. A pair of opera glasses, or keen eyesight, will show that Epsilon Lyrae is a wide double star with a separation of 208 seconds of arc (208″). However, closer examination with a telescope will reveal that each star is double again. Epsilon 1, the northernmost component, is comprised of 5th and 6th magnitude stars 2.8″ apart. It has been calculated that the two components of Epsilon 1 orbit each other over a period of around 1,165 years. The brightest of the two stars has been found to be a spectroscopic binary system. Epsilon 2, the southernmost pair, is made up of two 5th-magnitude stars orbiting each other over a period of 585 years. Most observers seem to agree that all the stars in the Epsilon system are white, although some differences of opinion have been voiced in the past, notably by William Herschel, who described the fainter component of Epsilon 1 as reddish. The relationship of Epsilon 1 to Epsilon 2 is uncertain. It may be that the two pairs are orbiting each other over a period of around a million years, although to date no orbital motion has been detected. It seems, therefore, that Epsilon 1 and Epsilon 2 form a physical double system.

Variable Stars

Many stars are seen to exhibit variations in brightness which take place on either a regular or irregular basis. These so-called variable stars are of two basic types; intrinsic variables are those whose brightness varies due to actual changes taking place within the star itself and extrinsic variables are those which vary because of the intervention of another object which could be either another star or clouds of gas and dust. There are also eruptive variables, which undergo sudden outbursts, brightening by several magnitudes, generally over fairly short periods of up to a few days, after which they return to their normal pre-outburst brightness.

Intrinsic Variables

There are a number of different classes of intrinsic variables, all of which are thought to be expanding and contracting, producing variations in their light output as they do so. Probably the best-known class is the *Cepheids*, the first to be discovered being Delta Cephei by the English astronomer John Goodricke in 1784. Cepheids are generally very luminous and have periods of anything between 1 and 50 days. Cepheids are interesting because they follow a distinct "period-luminosity relationship" whereby there is a definite link between their actual brightness and period. The longer the period, the more luminous the star. This relationship was discovered by the American astronomer Henrietta

Leavitt in 1912, and has since proved invaluable to astronomers as the distances to other galaxies can be determined. If the period of a Cepheid is measured, then its actual luminosity an be ascertained. Compare this with the observed brightness and it is fairly easy to assess the distance to the star.

Long-period variables are red giant stars whose brightnesses vary by several magnitudes over periods between 70 and 700 days. The periods of individual stars can alter slightly, as can their maximum and minimum magnitudes. The most famous example of a long-period variable is Omicron Ceti, better known as Mira, in the constellation of Cetus. Mira generally varies between 3rd and 9th magnitude over a period of 331 days or so, although it can occasionally attain 2nd magnitude. The variability of Mira was first noted by the Dutch astronomer David Fabricus in 1596.

Other intrinsic variables include *irregular variables*, such as Betelgeuse in Orion, which are giant stars whose brightness variation has no definite period, and *Beta Canis Majoris* stars, which are very bright stars that undergo slight variations over short periods.

Extrinsic Variables

This class includes both *eclipsing binaries* and *nebular variables*. In eclipsing binary systems there are two stars in orbit around each other, the plane of their orbit being in line with the Earth. Because of their orientation, the two stars are seen to alternately eclipse each other, thereby producing periodic reductions in the overall brightness of the system. Nebular variables vary due to clouds of gas and dust that surround the star. This material can sometimes obscure the light from the star, thereby producing irregular fluctuations in its brightness.

The most famous eclipsing binary is Algol, the variations of which were first noted by the Italian astronomer Geminiano Montanari in 1667. The Algol system consists of a hot blue-white star around which orbits a cooler red star. Every 2.87 days the fainter star passes in front of its brighter companion, reducing the overall magnitude from 2.1 to 3.4. There is a further, but smaller, reduction as the bright star eclipses its fainter neighbour. The presence of a third member of the Algol system has been verified spectroscopically. Named "Algol C", it has been found to orbit the eclipsing pair over a period of 680 days. A fourth star is also thought to revolve around the Algol system, although its presence has yet to be confirmed.

Typical of the nebular variables are the *T Tauri* stars. T Tauri itself lies at a distance of around 450 light years in the constellation of Taurus. It varies between magnitudes 10 and 10.5, although these fluctuations are totally unpredictable. It can be seen to lie quite close to the nebula NGC 1555, which is itself subject to unpredictable variations in brightness. T Tauri is thought to be a young star surrounded by clouds of gas and dust thrown off by the star as it attempts to settle down to a more stable existence.

Eruptive Variables

This class of variable embraces different types of nova-like stars. Novae are hot stars that increase greatly in brightness over a period of a few days before slowly fading back to normal. This gradual decrease in luminosity may take several years. One of the most spectacular novae of recent years was Nova Cygni, discovered by a Japanese amateur astronomer in the constellation of Cygnus in August 1975. Nova Cygni reached a magnitude of 1.8 and was in the order of a million times more luminous than the Sun. *Recurrent novae* are those which have been seen to brighten on more than one occasion, although their increases in luminosity are not as great as the novae, and they return to normal at a

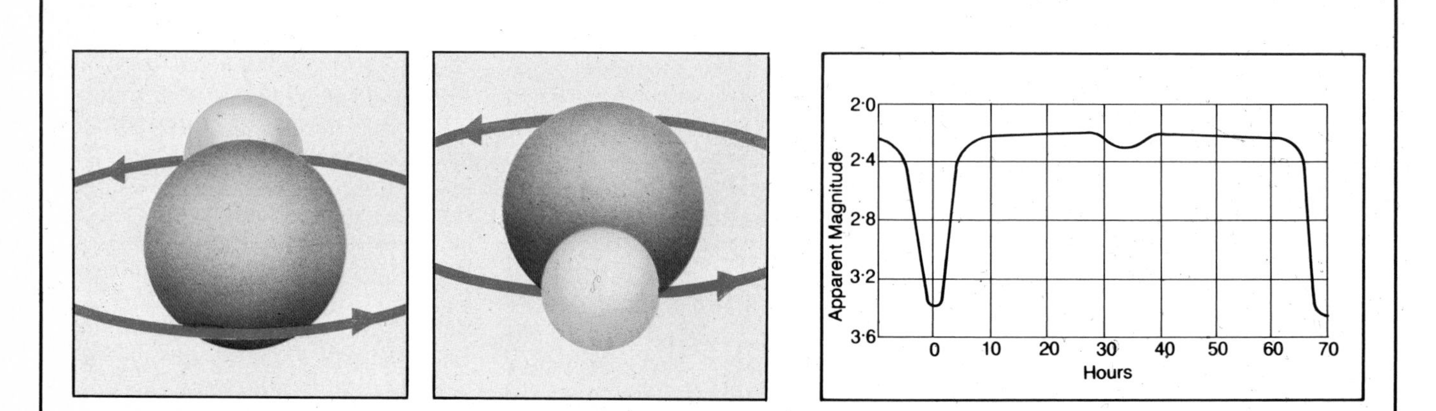

Above: *the variability of an eclipsing binary is caused by the orbital plane of the two stars being in line with the Earth. In the case of Algol, a fainter component passes in front of a brighter star producing a marked reduction in the overall light output of the system. At the opposite point in its orbit, the faint star passes behind its bright companion, resulting in only a slight reduction in brightness. As with all variable stars, the rise and fall in luminosity is plotted on a graph known as a light curve. In the light curve shown here, the large and small deviations in brightness are easily seen. They are shown against a magnitude scale on the vertical axis and a time scale along the horizontal axis.*

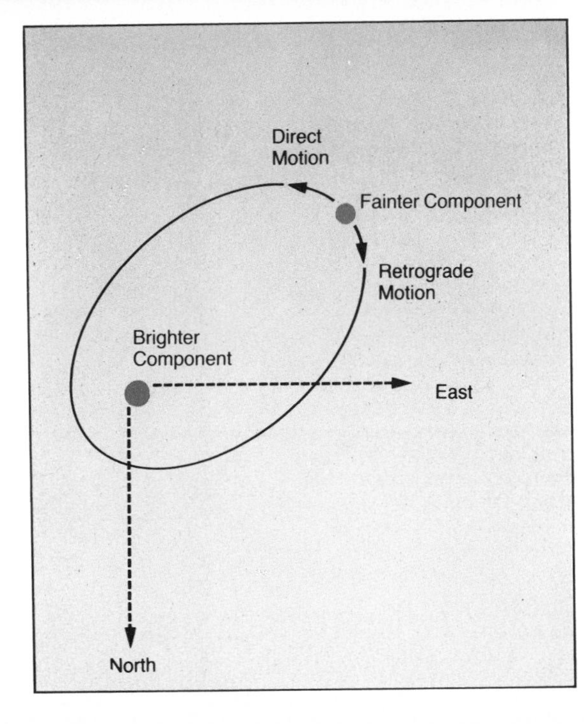

Above: *the constellation of Perseus with Algol seen to the right of picture.*
Right: *because the stars in binary systems orbit each other, their position angle and separation are constantly changing. If the orbit of the fainter component is producing an increase in PA, its motion is said to be direct, while a reduction is described as the fainter star having a retrograde motion. Any measurements of the positions of stars in a binary system are usually followed by an indication of the year in which the results were obtained. Due to these changes the members of many binary systems are drawing together and are becoming harder to see while others are becoming easier to resolve.*

much quicker rate.

Another type of eruptive variable are the *dwarf novae*, or *SS Cygni-type* stars. These stars undergo rapid increases in brightness and return to normal over a period of only a few weeks. This sequence is repeated every few months. SS Cygni itself brightens from magnitude 12 to 8 every 50 days or so.

Although there are many forms of nova, they all have one thing in common in that they are members of close binary systems that are thought to consist of a white dwarf and a larger, cooler star. Material is pulled from the cooler component and accumulates on the outer layers of the white dwarf. Eventually the temperature and pressure at the base of this new layer are sufficient to trigger off nuclear reactions, which eject the material into space and instigate a rapid but temporary increase in brightness.

Supernovae are very massive stars which explode after violent gravitational collapse (*see page 32*). Following virtual destruction they do not return to normal as do novae. Supernovae are fairly rare, and the last one to be seen in our Galaxy was that of 1604, discovered in the constellation of Ophiuchus.

Star Clusters

Although the majority of stars within our Galaxy are randomly distributed throughout space, many are seen to be members of small, relatively compact groups. A few of these groups, or clusters, can be seen with the naked eye, perhaps the best-known example of which is the Pleiades star cluster in the constellation of Taurus. Lying at a distance of around 400 light years, the stars within this cluster all formed from the same interstellar gas cloud, traces of which can still be seen throughout the cluster. Without a telescope only the six brightest members of the Pleiades can be seen, although telescopes reveal around 500 individual stars. There are two quite separate types of star cluster – *open* (or *galactic*) and *globular*.

The Pleiades is a typical open cluster and, like others of its type, lies in the galactic plane. There are over a thousand open clusters known to exist in our Galaxy alone, all of which orbit the galactic nucleus in more or less circular orbits. The number of stars in an open cluster can vary considerably, ranging from only 20 or so which form the rather sparse group M29 in Cygnus to the much larger Sword Handle Double Cluster in Perseus, which has a mass some 7,000 times that of the Sun.

In contrast to open clusters are globulars, which differ greatly in both their composition and distribution. Although open clusters have no well-defined shape, globular clusters, as their name suggests, are spherical formations that contain anything up to a million stars. Globulars visible to the naked eye appear as small, diffuse patches of light. The finest example is Omega Centauri in the constellation of Centaurus. It lies at a distance of around 17,000 light years and was catalogued by the Greek astronomer Ptolemy nearly 2,000 years ago. Omega Centauri is just one of some 200 globular clusters known to exist in the region of our Galaxy. Whereas there are very few open clusters with diameters of 30 light years, globular clusters can measure anything from 25 to 350 light years across. The stars within them are relatively tightly packed, which leads to a strong mutual gravitational attraction and it is because of this that they are thought to remain together over periods much longer than open clusters. It has been suggested, however, that strong gravitational perturbations can be set up which could push member stars either out into surrounding space or down towards the inner regions of the cluster. This may lead to a much denser gathering together of stars in the central regions, something which may be considered an ideal situation for the development of black holes. Observation has indeed revealed that quite a few globulars emit X-rays at such a level as to indicate black holes at the centres of these clusters.

Evolution of Stars in Star Clusters

Open clusters generally contain hot, blue-white stars which are similar in composition to our Sun. They are composed mainly of hydrogen, together with some helium and traces of certain

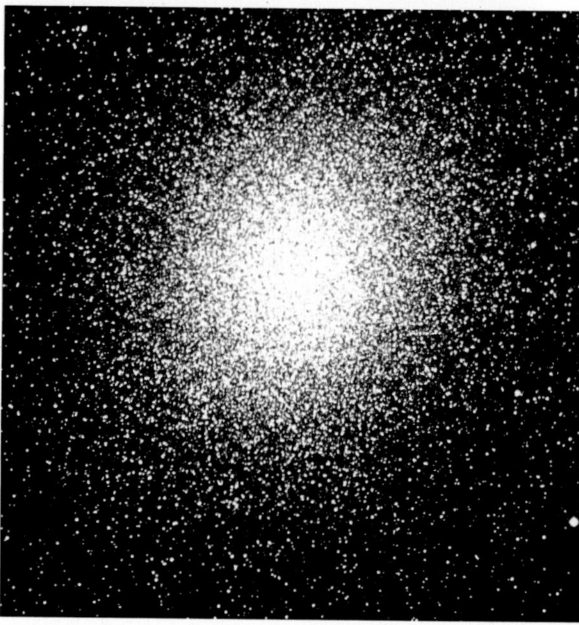

Above: *the Hyades and Pleiades open star clusters in Taurus.*
Above right: *the magnificent Sword Handle Double Cluster in Perseus.*
Right: *Omega Centauri, widely regarded as the most beautiful globular cluster in the sky.*
Middle right: *globular cluster 47 Tucanae.*
Far right: *open or galactic clusters* ***(top)*** *are composed of relatively young, hot stars of Population I and are found scattered along the main plane of the Galaxy. Globular clusters* ***(bottom)*** *are made up of older, cooler Population II stars and are found within the galactic halo, moving around the centre of the Galaxy in long, highly eccentric orbits.*

other, heavier elements. On the other hand, globular clusters are made up of much older red giant stars, typical of those found in galactic nuclei (*see Galaxies*). Also, because open clusters are much younger, many of them can still contain traces of the nebulosity remaining from the original gas cloud. The older globular clusters contain no such interstellar material.

Clusters in Other Galaxies

Open star clusters are collections of relatively young stars, and are known to exist throughout the Galaxy. Because their member stars are not as closely packed as in globulars the gravitational attraction between them is not as strong. As a result, open clusters can break up much more easily as stars escape into space. It is safe to assume that many of the individual stars we see in the night sky were once members of galactic clusters.

Globular clusters have been observed around other galaxies, notably the Andromeda Spiral in the constellation of Andromeda and the Sombrero Hat Galaxy in Virgo. The Andromeda Spiral has almost 200 globulars in attendance that lie at distances of up to 100,000 light years from the nucleus.

Quasars

In the early 1960s astronomers discovered a class of object that, although star-like in appearance, emitted vast amounts of energy. They were originally referred to as "quasi-stellar objects", a name which was eventually shortened to "quasars", and were found to be compact, extremely luminous and to lie at vast distances. The brightest quasar is 3c273 in the constellation of Virgo, which has been found to lie at the heart of a large elliptical galaxy. Other quasars have been identified as highly active galactic nuclei.

The most distant quasars currently known lie at a distance of around 15,000 million light years and their light has been travelling towards us since only a short time after the formation of the universe. We are therefore, in a sense, looking backwards in time, and it is through the study of quasars that we may extend our knowledge of the universe in which we live.

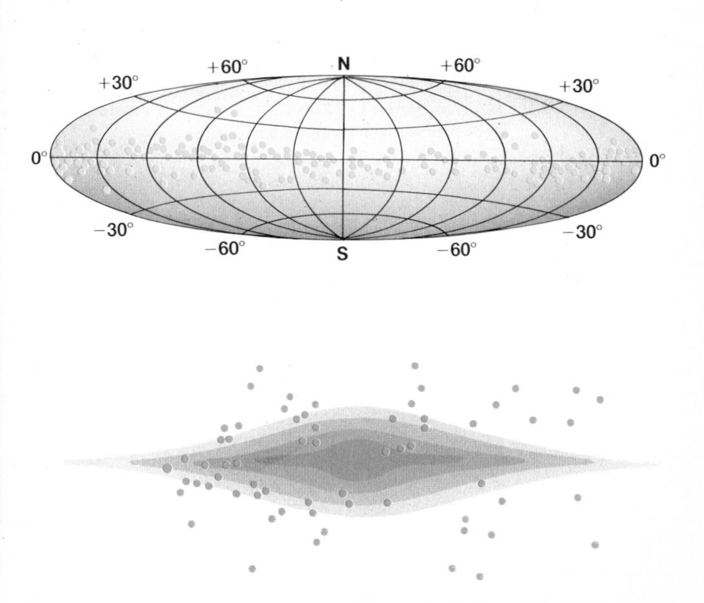

Nebulae

At various points in the sky, faint misty patches of light can be seen with the naked eye. Telescopes reveal that these objects are interstellar clouds of gas and dust. Known as nebulae, several hundred are known to exist within our Galaxy alone and others have been observed in neighbouring galaxies. Nebulae can appear either as bright, luminous clouds or as dark patches silhouetted against a brighter background. There are three main types of nebula; emission, reflection and dark.

Emission Nebulae

Emission nebulae contain hot stars which radiate energy at ultraviolet wavelengths. This energy is absorbed by the gas in the nebula, which in turn gives out visible light. One of the best examples is the Lagoon Nebula in Sagittarius, which has embedded within it a cluster of extremely young stars known as NGC 6530. These stars are thought to have formed only a couple of million years ago and the radiation they emit causes the gas surrounding the cluster to throw out visible light.

A telescopic view of the Lagoon will reveal two bright stars which are superimposed on the nebula. These stars are 9 Sagittarii and Herschel 36 and they also play an important role in causing the surrounding gas cloud to shine. The Lagoon Nebula has many other interesting features, including a dark dust lane which cuts through its centre. There are also a number of dark "globules" scattered across the foreground of the nebula. These are thought to be clumps of material which are in the process of collapsing to form new stars. The Lagoon Nebula lies at a distance from Earth of around 5,000 light years and is some 100 light years in diameter.

Reflection Nebulae

Reflection nebulae are generally less conspicuous than the emission type and they shine due to light from associated stars being reflected by dust particles within the cloud. The faint nebula M78 in Orion is a typical example. Located slightly to the north of the star Alnitak in the Belt of Orion, M78 is visible through small telescopes as a tiny, diffuse cloud surrounding a pair of 10th-magnitude stars. These stars provide the light by which M78 is seen to shine. Another reflection nebula is that enveloping the stars of the Pleiades cluster in Taurus. The Pleiades nebulosity is all that remains of the original cloud from which the stars in the cluster were formed, although it is believed that stars are still condensing in this region.

Dark Nebulae

Dark nebulae are clouds of matter that contain no stars and which blot out the light from brighter regions beyond. There are many examples, including the prominent Southern Coalsack in Crux Australis, the winding Snake Nebula in Ophiuchus and the famous Horsehead Nebula in Orion. The Horsehead was discovered in 1889 and was initially thought to be a gap in the bright emission nebula IC 434. However, we now know that IC 434 is merely a bright backdrop to the obscuring dust cloud we call the Horsehead Nebula. The Horsehead itself lies at a distance of 1,200 light years and is an extension of a much larger dust cloud immediately to the east.

Stellar Cast-offs

Two other types of nebula can be seen in our skies. Planetary nebulae and supernova remnants are the discarded outer layers of ageing stars. Planetary nebulae can be seen as shells of gas surrounding hot central suns, while supernova remnants bear testament to violent explosions of very massive stars approaching the final stages of evolution. Planetary nebulae are commonplace and their numbers include the Dumbbell Nebula in Vulpecula and the Helix Nebula in Aquarius. Supernova remnants can vary greatly in shape and size, ranging from the relatively compact Crab Nebula in Taurus, which arose from a stellar explosion witnessed less than a thousand years ago, to the vast 100-light year diameter Veil Nebula in Cygnus. The supernova which formed the Veil Nebula blazed out over 100,000 years ago, and its remains now take the form of a gigantic shell of gas that is slowly expanding and dissipating into surrounding space.

Opposite top left: *the bright, diffuse Lagoon Nebula in Sagittarius.*
Opposite left: *the stars in the Pleiades cluster are seen to be wreathed in nebulosity, remnants of the original cloud from which the stars were formed.*
Above left: *the Crab Nebula in Taurus represents the remnants of a supernova witnessed in AD1054.*
Above: *the well-known Horsehead Nebula in Orion.*
Left: *The Veil Nebula, a faint supernova remnant in Cygnus.*

Galaxies

The Milky Way

Our Sun is one of around 100,000 million stars within the Milky Way Galaxy. Together with these stars the Milky Way is home to vast amounts of interstellar gas and dust from which new stars are continually being formed. The Milky Way is a typical spiral galaxy and has three main regions. The *central bulge* contains old, relatively cool Population II stars but very little gas and dust. The *disc* contains the spiral arms, which are made up of younger Population I stars together with a much higher percentage of interstellar material. These Population I stars are generally much hotter than those found in the central regions. There are a total of four arms, which are, from the centre, the Norma Arm, the Sagittarius Arm, the Orion Arm and the Perseus Arm. The Sun is situated some 30,000 light years from the centre, roughly two-thirds of the way out, between the Orion and Perseus Arms. The *halo* is seen to surround the central region of the Galaxy and again contains very old stars. These stars are concentrated into globular clusters, which move around the galactic centre in highly elongated orbits.

As we look along the plane of the Galaxy the effect we see is that of countless thousands of faint stars, which, although too dim to be seen individually, combine to produce a misty band of light which stretches across the sky. This is the Milky Way and, although its presence has been known ever since man first looked at the night sky, its composition remained a mystery until Galileo first turned his telescope towards it in the early part of the seventeenth century. The centre of the Galaxy lies in the direction of the constellation Sagittarius, although we are unable to see it due to the presence of vast obscuring clouds of dust situated throughout the inner regions of the galactic disc.

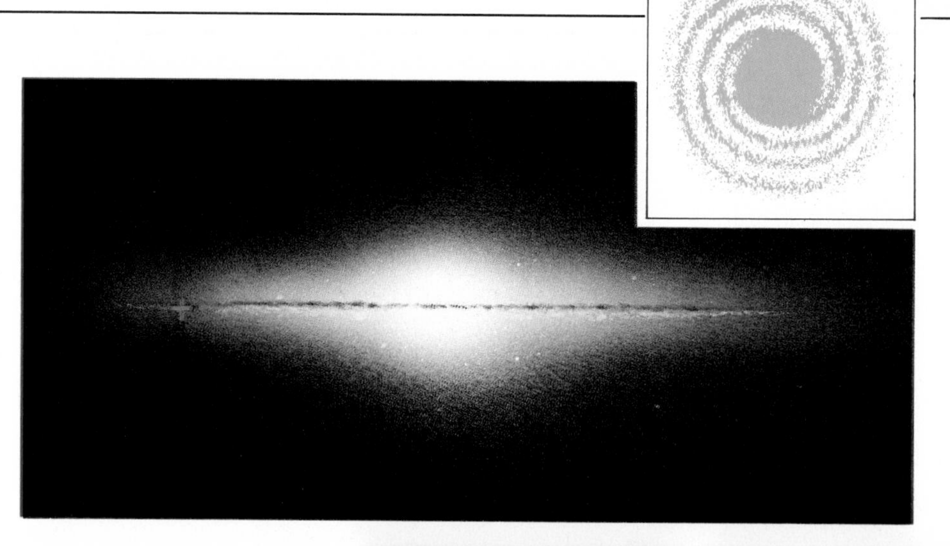

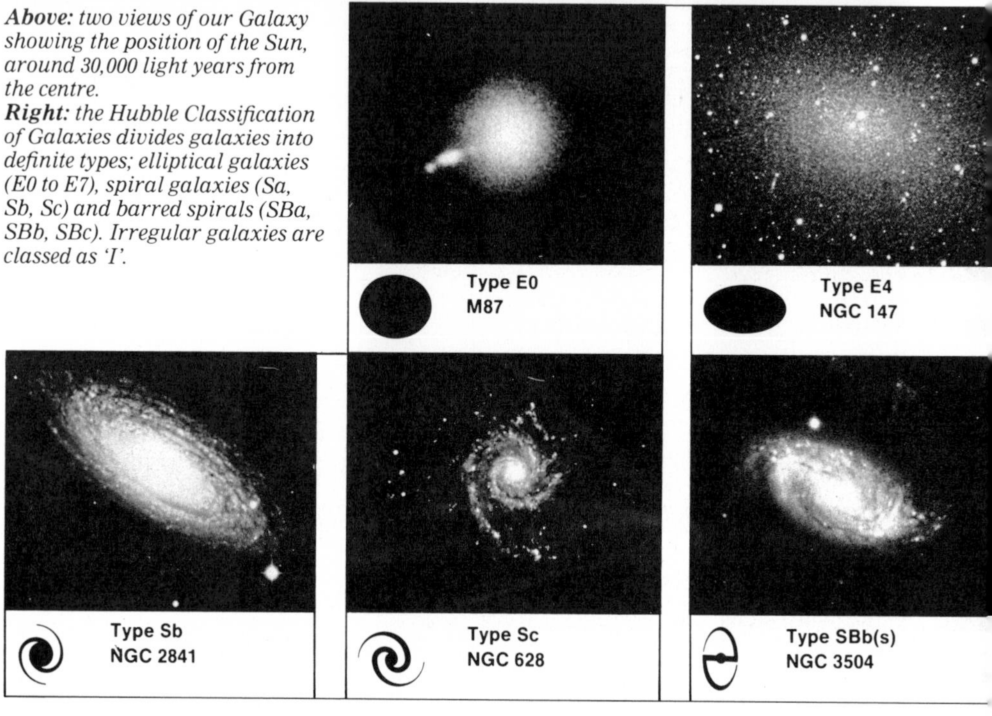

Above: *two views of our Galaxy showing the position of the Sun, around 30,000 light years from the centre.*
Right: *the Hubble Classification of Galaxies divides galaxies into definite types; elliptical galaxies (E0 to E7), spiral galaxies (Sa, Sb, Sc) and barred spirals (SBa, SBb, SBc). Irregular galaxies are classed as 'I'.*

Types of galaxies

The vast majority of known galaxies can be divided into two basic types; elliptical and spiral. Elliptical galaxies, as their name suggests, are uniform in appearance and can vary from being almost spherical to very elongated. Spiral galaxies have a central region containing the nucleus. Surrounding this is the disc, in which spiral arms can be seen. In ordinary spirals these arms emerge from the central regions, although in barred spirals the arms emanate from each end of a bar that crosses the galactic plane. In all spirals the arms trail behind as the galaxy rotates around its nucleus. Both elliptical and spiral galaxies are classified according to a system first proposed by the American astronomer Edwin Hubble in 1925.

Other types of galaxy include irregular galaxies, in which there is no well-defined shape; lenticular galaxies, which have a central region, although the surrounding disc displays no evidence of spiral arms; radio galaxies, which are very strong sources of radio emission, and Seyfert galaxies, first noted by the American astronomer Carl Seyfert in 1943, which have weak spiral arms surrounding a small but bright nucleus. Of these latter four types, only the lecticular galaxies fit into the Hubble system, falling between the elliptical and spiral types.

Groups of galaxies

The Milky Way Galaxy is a member of the Local Group, a cluster of around 30 spiral, elliptical and irregular galaxies. Many of these galaxies are concentrated around the two dominant

Left: *edge-on spiral galaxy Messier 82 in Ursa Major.*
Below: *the Andromeda Spiral galaxy.*

members of the Local Group; our own Galaxy and the Andromeda Spiral. The giant Andromeda Spiral is, in fact, the largest member of the Local Group. It has a diameter in the region of 150,000 light years and contains examples of every class of object found within our own Galaxy, including double and variable stars, star clusters, nebulae and so on. Apart from these two galaxies, and the Triangulum Spiral, the Local Group is comprised of dwarf elliptical and irregular galaxies.

By comparison the Local Group is quite small and clusters of several thousand galaxies are known to exist. Within a distance of 50 million light years over 50 clusters have been observed, including those in Sculptor, Cetus, Ursa Major and Fornax. They range in size from small systems like our own to the gigantic Virgo Cluster, which contains over 1,000 galaxies. It is generally believed that many of the clusters we see are actually members of enormous clusters of clusters, or "superclusters", the Local Group being part of one such supercluster centred on the Virgo Cluster of galaxies.

Expansion of the universe

The general concensus of opinion among astronomers is that the material we see scattered throughout the universe was originally concentrated into a single, super-dense mass, and that the universe came into being around 15,000 million years ago as a result of the release of this material in a colossal explosion commonly referred to as the Big Bang. Much of the observational evidence we have supports the Big Bang theory. For example, it has been shown that clusters of galaxies are moving apart in a motion that is thought to stem from such an event. A universal background radiation has also been detected that is believed to be a relic of the original Big Bang explosion. Other theories have been suggested on the birth of the universe, but these have generally been abandoned.

The Northern Sky: Winter

The northern winter sky sees Cassiopeia high in the north-west. The Plough, formed from the seven brightest stars in Ursa Major, is prominent high in the east. Ursa Minor, with Polaris, is almost surrounded by Draco. High overhead, Auriga and its brilliant star Capella can be seen, while Pegasus and Andromeda are setting in the west. Looking towards the south, the unmistakable form of Orion strides across the sky followed by his two faithful dogs – Canis Major and Canis Minor – as he relentlessly pursues Taurus the Bull. Creeping into the south-eastern sky is Leo with his promise of warmer nights.

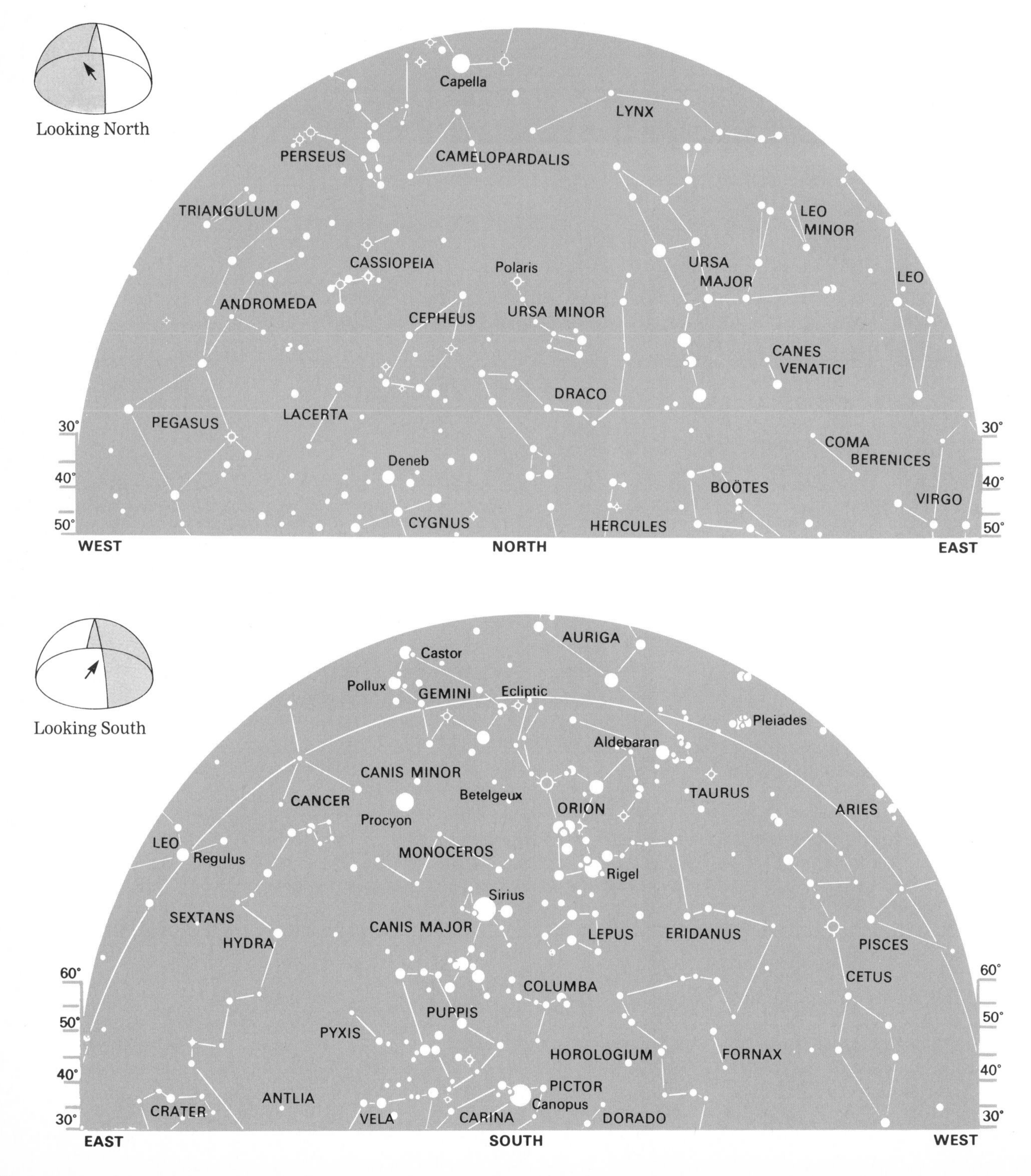

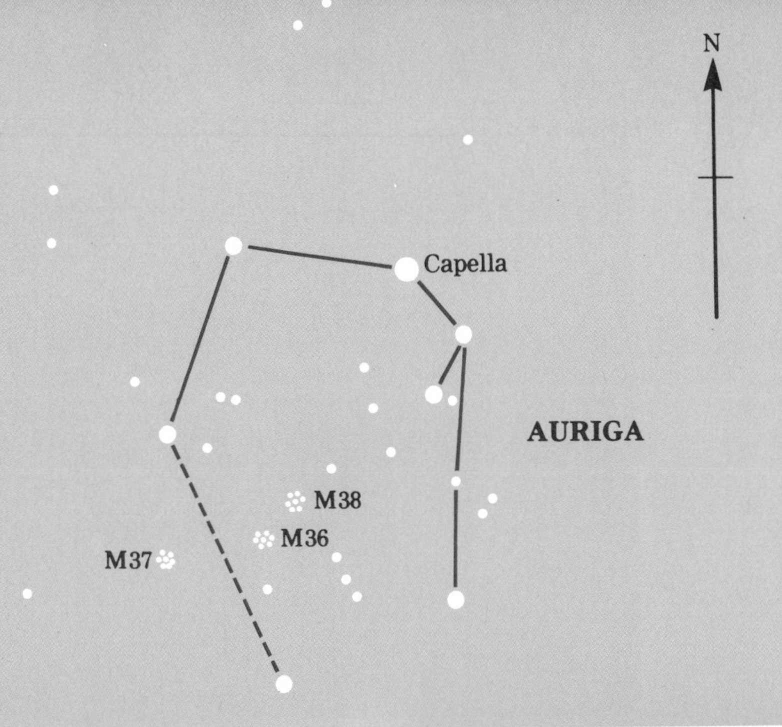

The brightest star in Auriga is golden-yellow Capella, which shines from a distance of 45 light years. Its magnitude of 0.06 makes it the 6th brightest star in the sky and measurements show that its actual luminosity is over 150 times that of our Sun. Auriga contains three bright open star clusters; M36, M37 and M38. All three are visible in binoculars although even a small telescope will resolve individual stars in each, particularly M37, considered by many to be the finest of the three. M36 contains around 60 stars, while M38, situated a little to the north-west, has over 100 members. M37 is by far the largest with a total of well over 150 suns.

Canis Minor is a very small constellation. Its brightest member is Procyon, a white star which lies at a distance of just over 11 light years. 14 Canis Minoris has two faint stars quite close to it; all three are visible through binoculars. These stars are not actually related to each other, but happen to lie in the same line of sight.

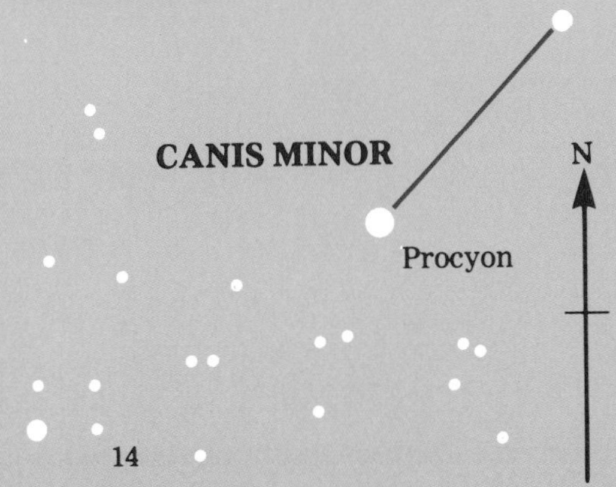

Right: *the Crab Nebula is a supernova remnant and is believed to represent the scattered remains of a star which was seen to explode in AD 1054. It lies at a distance of around 6,000 light years and is found a little to the north-west of Zeta Tauri. At 9th magnitude it can just be seen with binoculars if the sky is really dark and clear.*

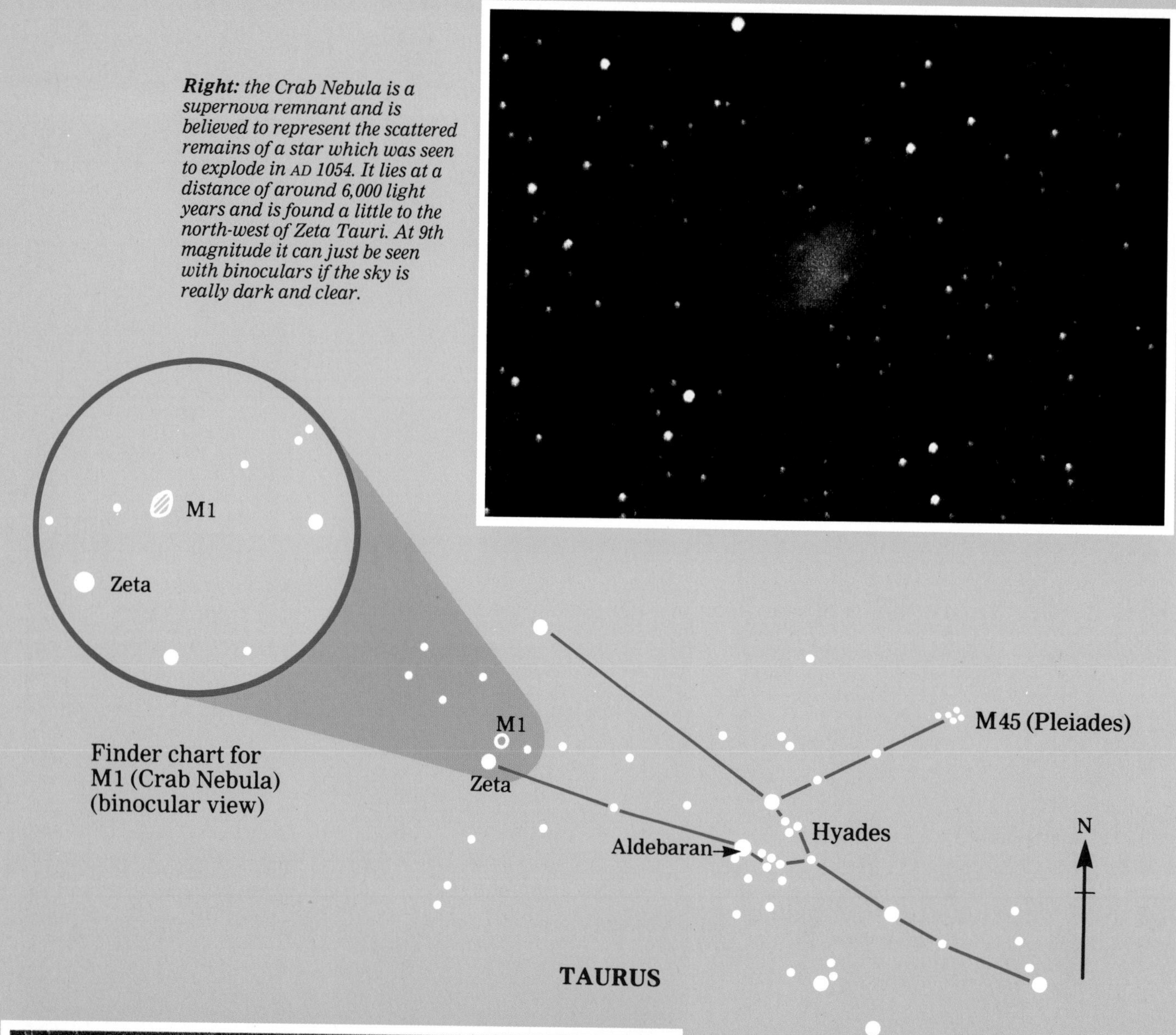

Finder chart for M1 (Crab Nebula) (binocular view)

Left: *this is a closer view of the Hyades cluster. This group forms the Head of Taurus the Bull and contains several hundred stars, well over a hundred of these being visible through binoculars. The cluster is quite large with outlying members scattered over a wide area. The V-shaped gathering shown here is around 8 light years across and represents only the central region of the Hyades. Also shown is the fainter and more distant cluster NGC 1647, seen near the left side of the picture. NGC 1647 contains around 30 stars and shines with a magnitude of 6.3. It can be spotted quite easily with binoculars or a small telescope.*

Right: *the Pleiades is without doubt the most beautiful object of its kind in the sky. The cluster lies at a distance of just over 400 light years and on average around 6 or 7 member stars can be seen with the naked eye, the brightest of which is Alcyone, seen just left of centre. The stars in the Pleiades are shrouded in reflection nebulosity, although this is difficult to see without a large telescope. As with all open star clusters the best views are obtained either with binoculars or a wide-field telescope.*
Below: *this view shows both the Pleiades and the V-shaped Hyades, the latter being one of the nearest open clusters to us, lying at a distance of only 130 light years. The bright star at lower left of the group is Aldebaran. This isn't actually a cluster member but lies in the same line of sight at just over half the distance.*

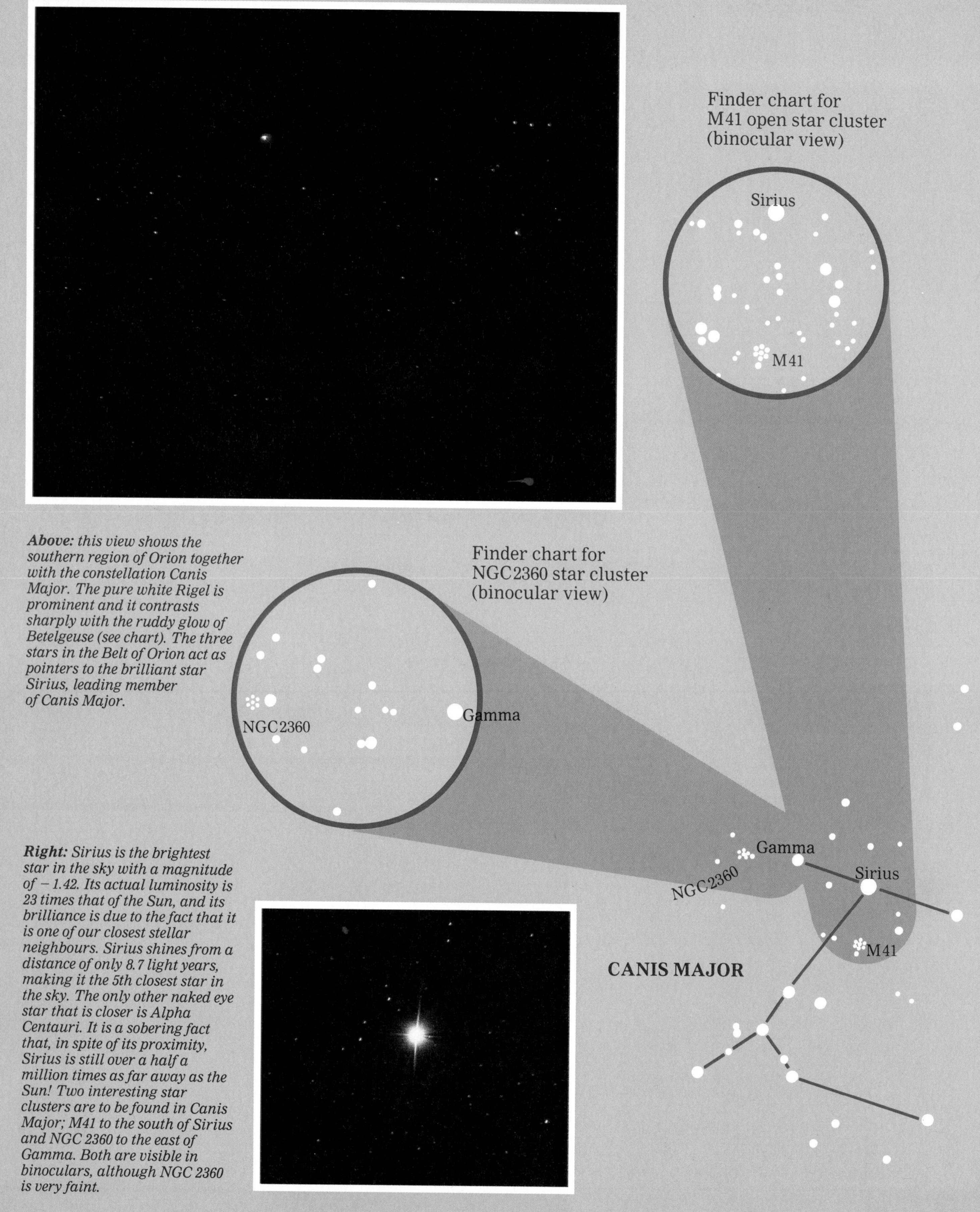

Above: *this view shows the southern region of Orion together with the constellation Canis Major. The pure white Rigel is prominent and it contrasts sharply with the ruddy glow of Betelgeuse (see chart). The three stars in the Belt of Orion act as pointers to the brilliant star Sirius, leading member of Canis Major.*

Right: *Sirius is the brightest star in the sky with a magnitude of –1.42. Its actual luminosity is 23 times that of the Sun, and its brilliance is due to the fact that it is one of our closest stellar neighbours. Sirius shines from a distance of only 8.7 light years, making it the 5th closest star in the sky. The only other naked eye star that is closer is Alpha Centauri. It is a sobering fact that, in spite of its proximity, Sirius is still over a half a million times as far away as the Sun! Two interesting star clusters are to be found in Canis Major; M41 to the south of Sirius and NGC 2360 to the east of Gamma. Both are visible in binoculars, although NGC 2360 is very faint.*

***Right:** this view shows the Belt of Orion together with his Sword and the Orion Nebula, seen at the bottom of the picture. Delta Orionis, also known as Mintaka, has a companion star of magnitude 6.7 situated 52.8 seconds of arc away to the north. The Belt of Orion is worth sweeping with binoculars. Another double is Lambda Orionis, which has two components of magnitudes 3.7 and 5.6 lying 4.4 seconds of arc apart.*

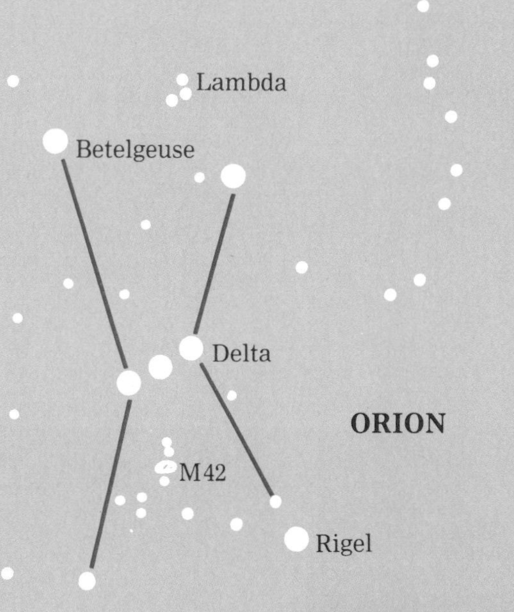

***Left:** the Orion Nebula is regarded by many as the finest nebula in the heavens and is a favourite for observers of the winter sky. Small telescopes reveal a bright greenish patch of light shimmering against the dark background of sky. Hot stars within the nebula emit ultraviolet radiation. The gas in the nebula absorbs this radiation and in turn emits the visible light we see. The Orion Nebula is known to be a birthplace of stars and the formation of new suns is currently under way deep inside this misty patch of fluorescent gas in the Sword of Orion.*

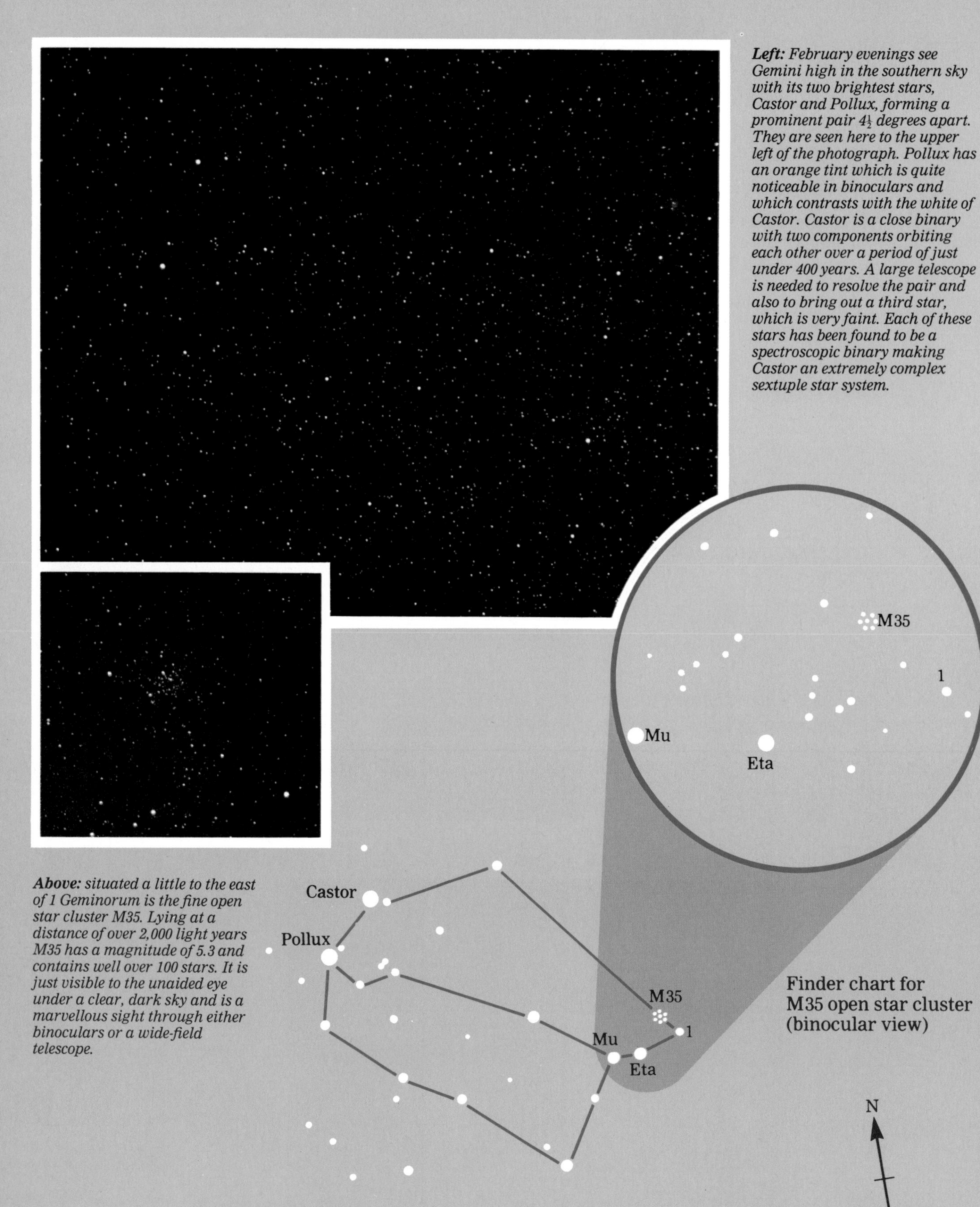

Left: *February evenings see Gemini high in the southern sky with its two brightest stars, Castor and Pollux, forming a prominent pair 4½ degrees apart. They are seen here to the upper left of the photograph. Pollux has an orange tint which is quite noticeable in binoculars and which contrasts with the white of Castor. Castor is a close binary with two components orbiting each other over a period of just under 400 years. A large telescope is needed to resolve the pair and also to bring out a third star, which is very faint. Each of these stars has been found to be a spectroscopic binary making Castor an extremely complex sextuple star system.*

Above: *situated a little to the east of 1 Geminorum is the fine open star cluster M35. Lying at a distance of over 2,000 light years M35 has a magnitude of 5.3 and contains well over 100 stars. It is just visible to the unaided eye under a clear, dark sky and is a marvellous sight through either binoculars or a wide-field telescope.*

Situated to the east of Orion, the constellation of Monoceros is fairly large but faint and somewhat obscure. However, the Milky Way passes through this region and the area is well worth sweeping with binoculars, which will bring out some very pretty star fields. The open cluster M50 can be found to the south of 19 and 20 Monocerotis. This is a fine gathering of stars and at a magnitude of 6.3 is an easy object for binoculars. M50 lies at a distance of around 3,000 light years. There is a single red giant star just to the south of the centre of the cluster which stands out quite well against the blue and white stars that form the rest of M50.

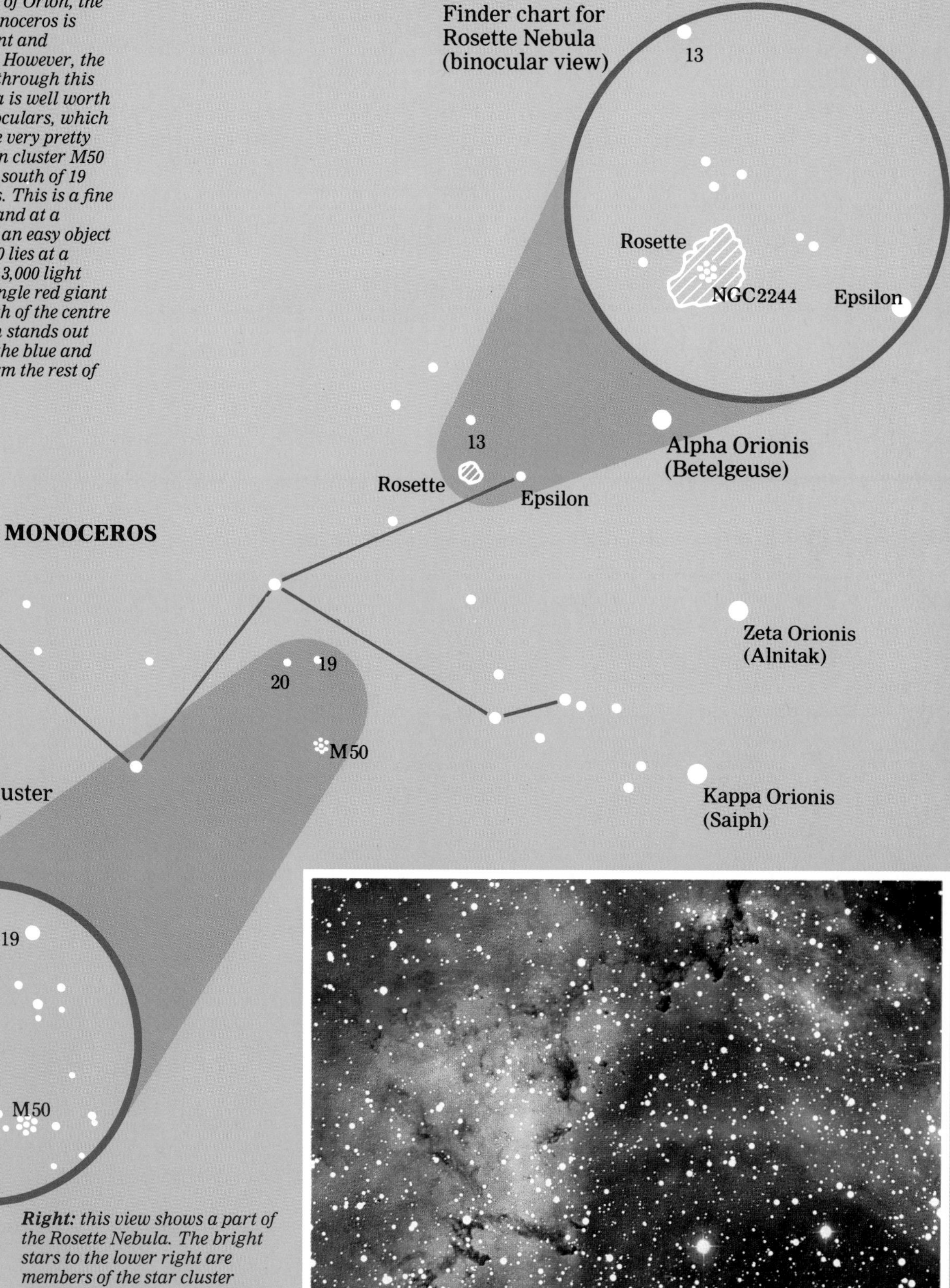

Right: *this view shows a part of the Rosette Nebula. The bright stars to the lower right are members of the star cluster NGC 2244, which is embedded within the nebula. The Rosette is an emission nebula and a stellar birthplace. Binoculars will show the 6th magnitude cluster quite well, although the nebula itself is quite faint and will only reveal itself under ideal conditions.*

The Northern Sky: Spring

The procession of spring constellations is led by Leo, which lies high in the southern sky. The Great Bear is almost overhead and if the curve formed by the stars in the Bear's tail is followed down towards the southeast it points the way to the conspicuous reddish yellow Arcturus in Boötes and the brilliant white Spica in Virgo. The "W" of Cassiopeia is low in the north and Orion is chasing away the cold winter nights as he follows Taurus towards the western horizon. One of the brightest stars in the sky is the yellow giant Capella, shining like a cosmic searchlight high up in the northwest.

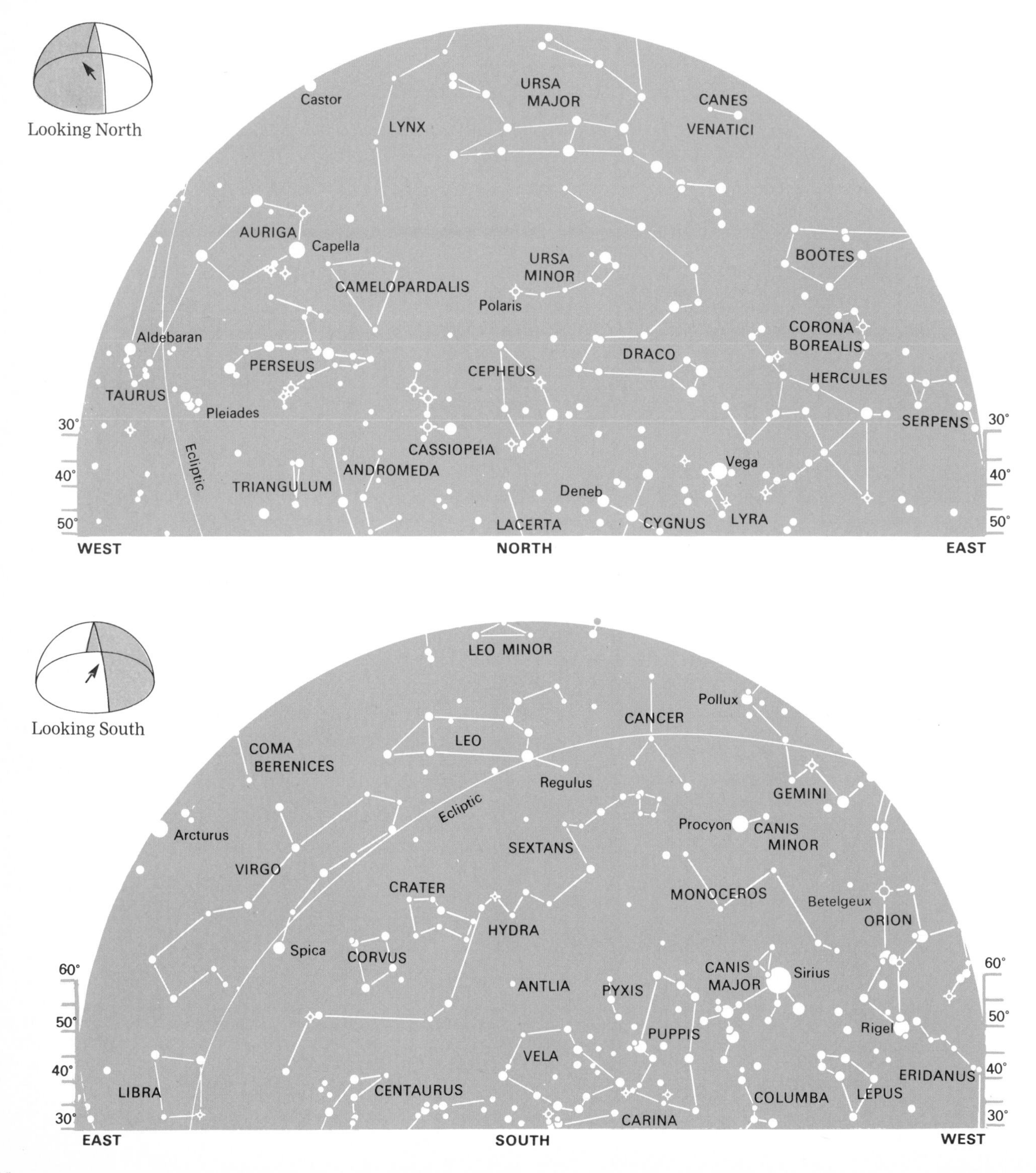

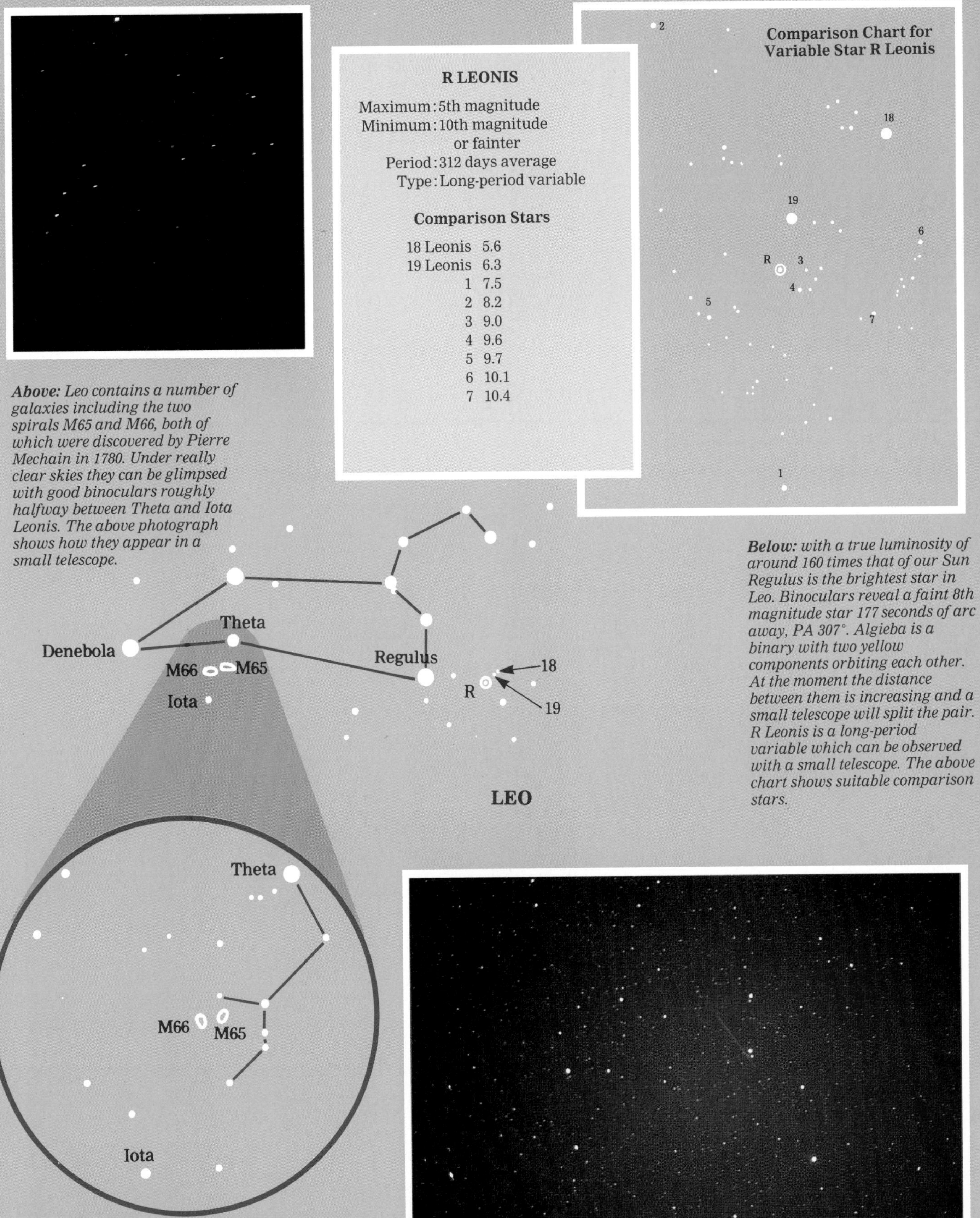

R LEONIS

Maximum: 5th magnitude
Minimum: 10th magnitude or fainter
Period: 312 days average
Type: Long-period variable

Comparison Stars

Star	Magnitude
18 Leonis	5.6
19 Leonis	6.3
1	7.5
2	8.2
3	9.0
4	9.6
5	9.7
6	10.1
7	10.4

Above: *Leo contains a number of galaxies including the two spirals M65 and M66, both of which were discovered by Pierre Mechain in 1780. Under really clear skies they can be glimpsed with good binoculars roughly halfway between Theta and Iota Leonis. The above photograph shows how they appear in a small telescope.*

Below: *with a true luminosity of around 160 times that of our Sun Regulus is the brightest star in Leo. Binoculars reveal a faint 8th magnitude star 177 seconds of arc away, PA 307°. Algieba is a binary with two yellow components orbiting each other. At the moment the distance between them is increasing and a small telescope will split the pair. R Leonis is a long-period variable which can be observed with a small telescope. The above chart shows suitable comparison stars.*

Finder chart for spiral galaxies M65 and M66

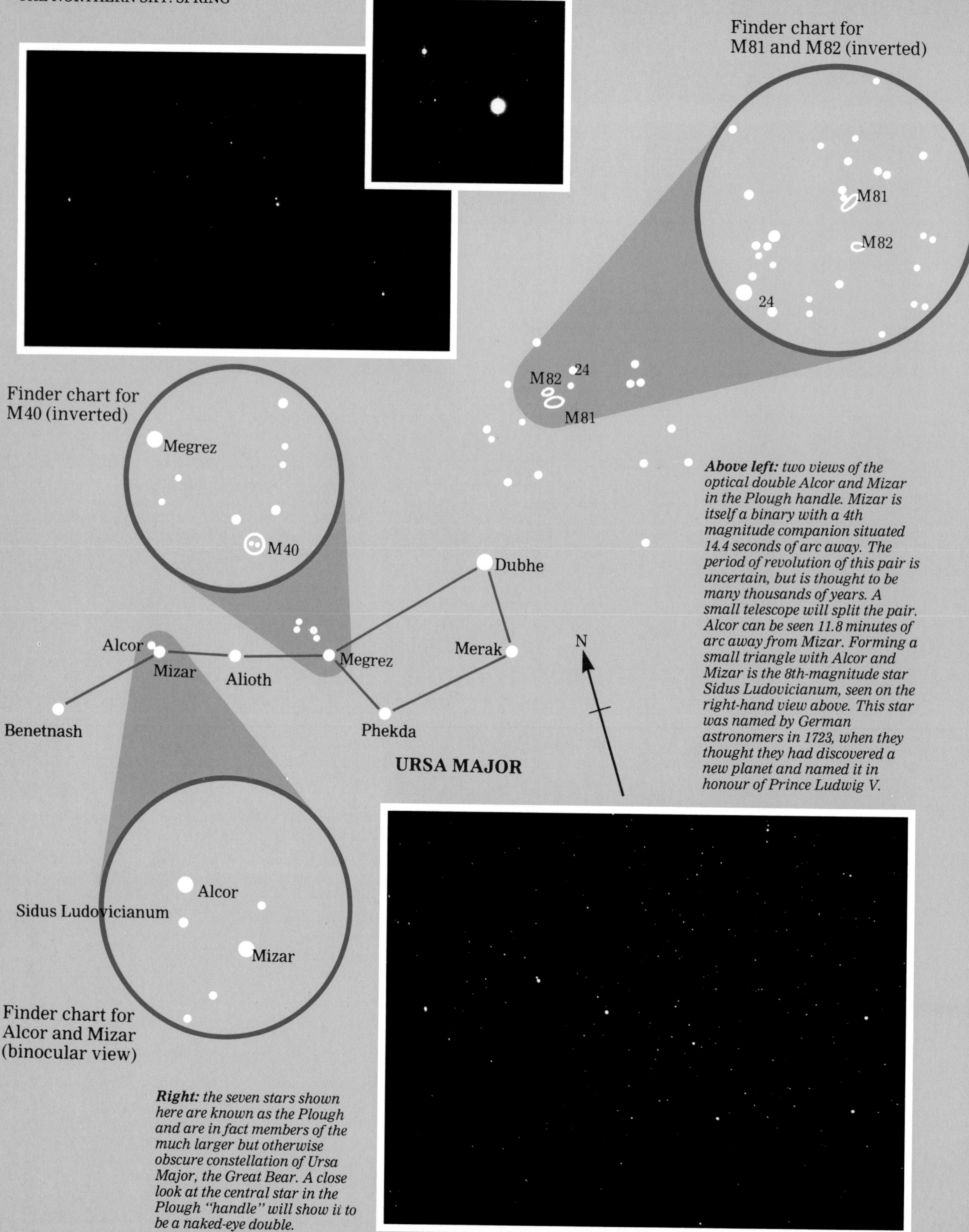

Above left: *two views of the optical double Alcor and Mizar in the Plough handle. Mizar is itself a binary with a 4th magnitude companion situated 14.4 seconds of arc away. The period of revolution of this pair is uncertain, but is thought to be many thousands of years. A small telescope will split the pair. Alcor can be seen 11.8 minutes of arc away from Mizar. Forming a small triangle with Alcor and Mizar is the 8th-magnitude star Sidus Ludovicianum, seen on the right-hand view above. This star was named by German astronomers in 1723, when they thought they had discovered a new planet and named it in honour of Prince Ludwig V.*

Right: *the seven stars shown here are known as the Plough and are in fact members of the much larger but otherwise obscure constellation of Ursa Major, the Great Bear. A close look at the central star in the Plough "handle" will show it to be a naked-eye double.*

Left: *view of the area of sky around M81 and M82 showing the two galaxies as they appear in a telescope. M81, towards the upper left, is a fine example of a spiral galaxy. At magnitude 7.9 it is somewhat brighter than M82, which shines with a magnitude of 8.8. M82, seen towards the lower right of the photograph, is also a spiral but seen edge-on. Both galaxies are members of a small cluster of galaxies situated at a distance of around 7 million light years.*
Above: *a closer view of M82, which appears as an elongated patch of light crossed with numerous dark lanes.*

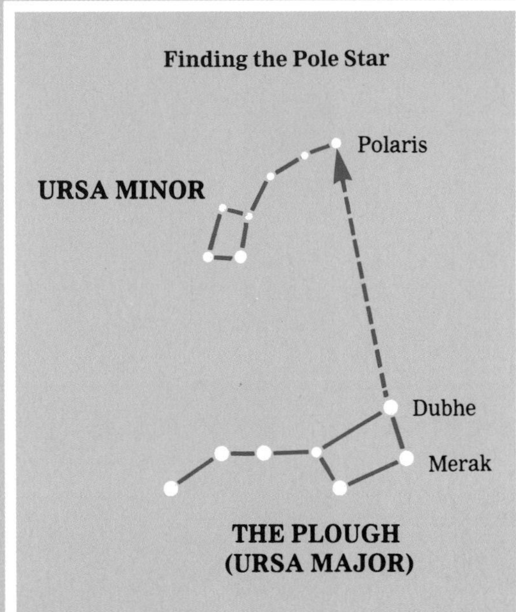

Right: *the Pole Star can be located quite easily by using the two end stars in the "bowl" of the Plough as pointers. A line from Merak, through Dubhe, extended for a further 25 degrees or so will lead to Polaris, a moderately bright star situated in an otherwise barren field. The rest of Ursa Minor, the Little Bear, can be seen stretching away from Polaris.*

Left: *the famous Whirlpool Galaxy is an 8th magnitude face-on spiral situated just to the southwest of Benetnash, the star at the end of the Plough handle. The galaxy can be seen in the same field as 24 Canum Venaticorum. Also known as M51, the Whirlpool Galaxy was discovered by Charles Messier in 1773 and lies at a distance of around 35 million light years. It is just about visible in binoculars on a clear, dark night, although large telescopes are required to bring out any significant details. Long-exposure photographs made with large telescopes reveal individual stars, star fields, bright concentrations of material that may be star clusters, and bright and dark nebulosity. M51 has a diameter of about 100,000 light years, making it comparable in size to the Andromeda Spiral.*

Right: *the globular cluster M3 is one of the brightest globulars in the sky and shines with a magnitude of 6.4. It is quite easy to find, situated to the south of 25 Canum Venaticorum, and appears as a bright nebulous spot in binoculars. In order to resolve any individual stars, however, a telescope of at least 6 inches aperture will be required. M3 lies at a distance of almost 40,000 light years and is thought to contain upwards of half a million stars.*

*The photograph at **lower right** shows most of Canes Venatici, including the area around the spiral galaxy M94. Forming a triangle with 9 and 10 Canum Venaticorum, M94 is a fairly bright object, shining with a magnitude of 7.9, and can be found quite easily in a wide-field telescope. Under ideal seeing conditions, binoculars may also bring it out. If you don't locate this, or indeed any other faint object, straightaway, then don't worry. The more time you spend actually looking at the sky the easier it will become to pick out faint nebulous objects. Astronomical observing is something that becomes easier, and more rewarding, with practice. The area of sky around Canes Venatici is rich in faint galaxies and is well worth sweeping with a large telescope.*

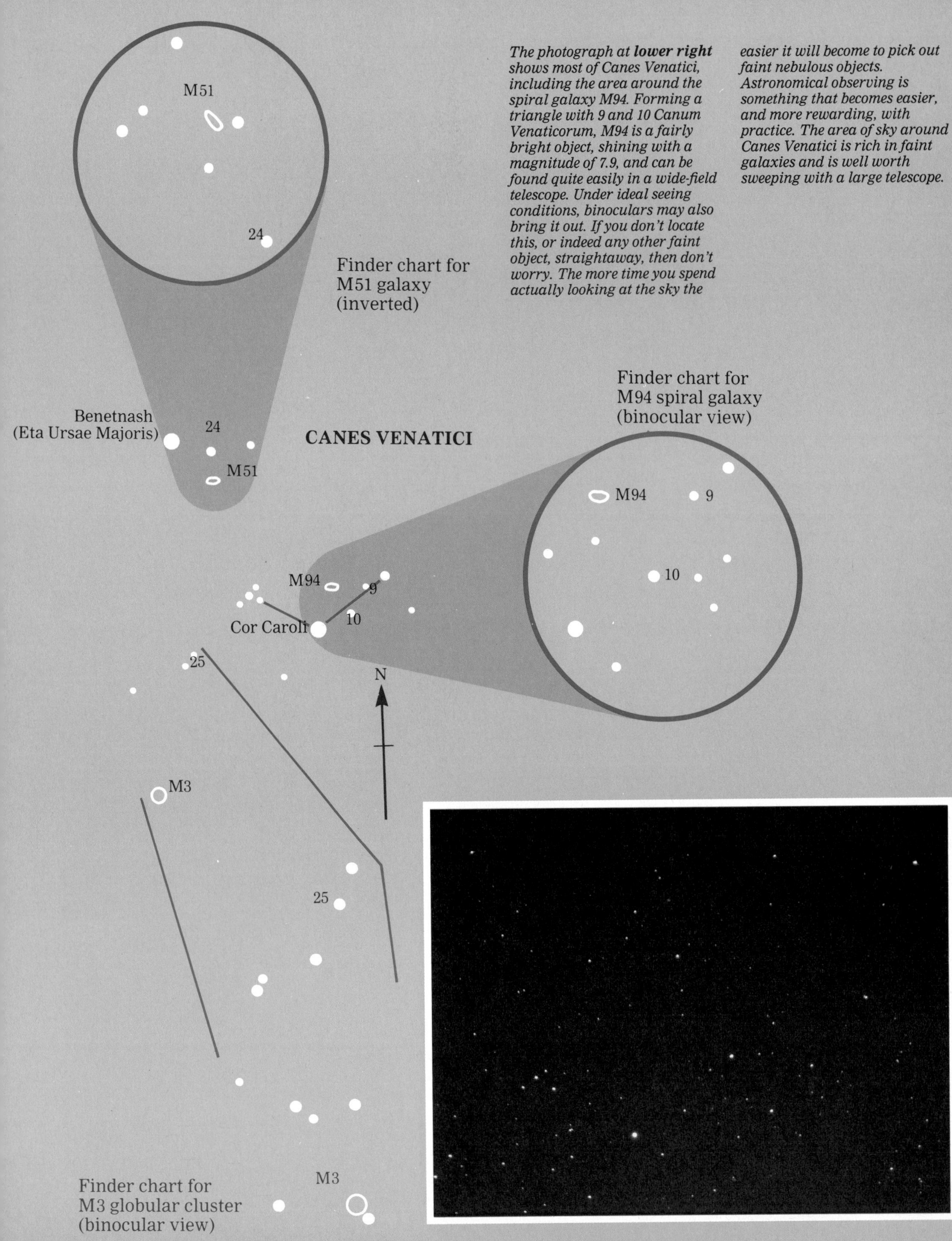

Finder chart for M51 galaxy (inverted)

Finder chart for M94 spiral galaxy (binocular view)

Finder chart for M3 globular cluster (binocular view)

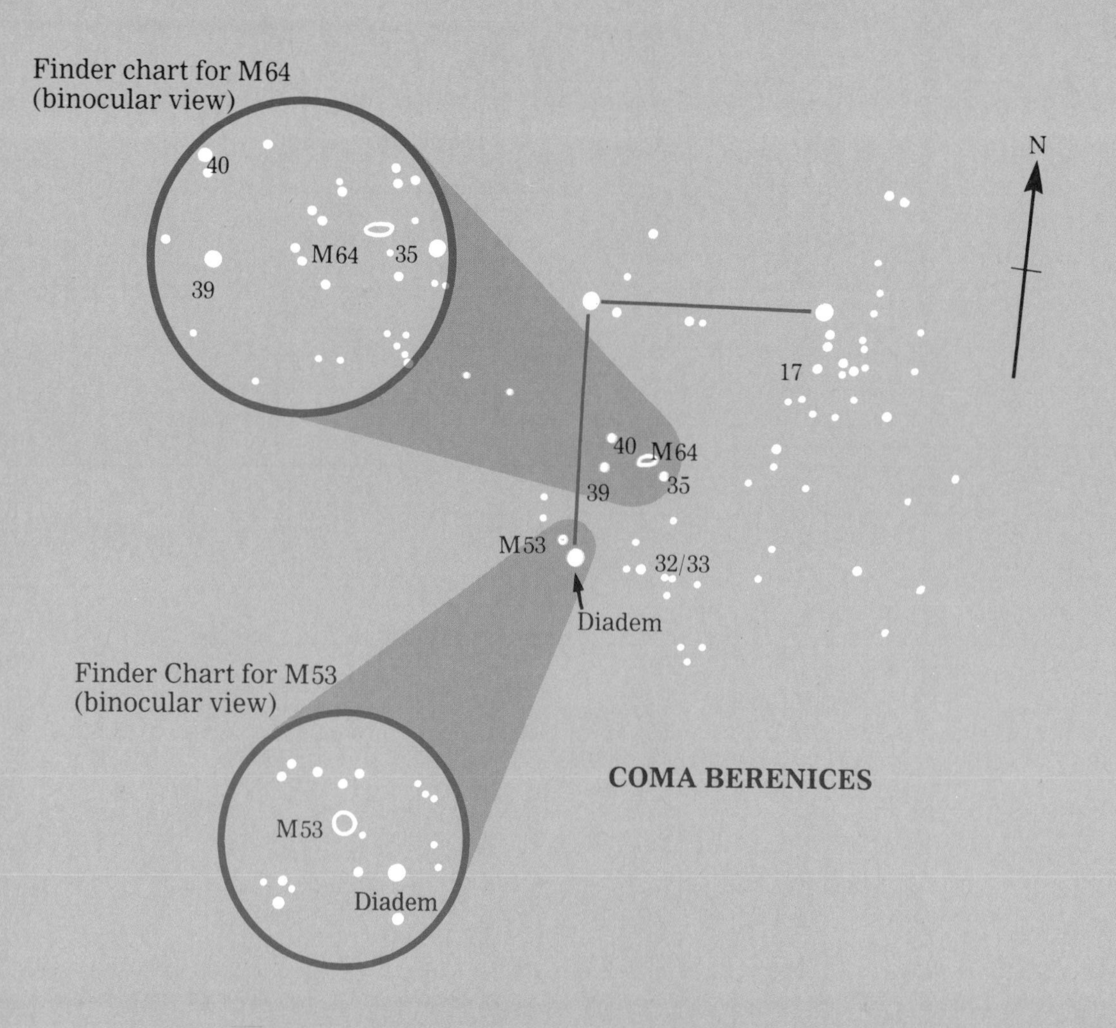

Right: *a good pair of binoculars will show M64, the famous "Black Eye Galaxy". This is an 8th magnitude spiral system situated inside the triangle of stars 35, 39 and 40 Comae Berenicis, less than a degree away from 35. M64 lies at a distance of over 20 million light years and can be seen to have a conspicuous swathe of dust stretching across its central regions. This feature has given rise to the popular name applied to the galaxy.*

Left: *this photograph shows most of Coma Berenices, which, to the naked eye, takes on the appearance of a large and sparsely populated open star cluster. There are a number of objects worth seeking out, including the globular cluster M53, situated in the same low-power field as Alpha Comae Berenicis. M53 can be seen in binoculars as a faint, circular patch of light shining with a magnitude of 7.6. To the west of Alpha is the binocular double comprised of 32 and 33, distance 195 seconds of arc, PA 49°. The area of Coma Berenices plays host to many faint galaxies, a large number of which can be observed through medium-sized telescopes.*

Right: *many of the stars in Coma Berenices are actually members of a very loose gathering known as the Coma Star Cluster. This is one of the closest open star clusters and lies at a distance of around 250 light years. One of the stars in the cluster is 17 Comae Berenicis. This is a double with components of magnitudes 5.4 and 6.7, distance 145 seconds of arc, PA 251°. Binoculars will show the pair well. In all, the Coma Star Cluster has almost 40 members contained within an area of sky roughly 5° in diameter.*

The Northern Sky: Summer

The summer sky is dominated by the Summer Triangle, a conspicuous configuration of stars formed from the brilliant blue-white Vega in Lyra and Deneb in Cygnus, both of which are almost overhead. Further to the south, Altair in Aquila makes up the trio. The Great Bear is over the north-western horizon, slowly descending to his autumnal resting place low in the north. One of the brightest summer stars is Antares in Scorpius, quite unmistakable with its strong red hue. To the east of Scorpius the beautiful star clouds of Sagittarius beckon the attentions of the backyard astronomer.

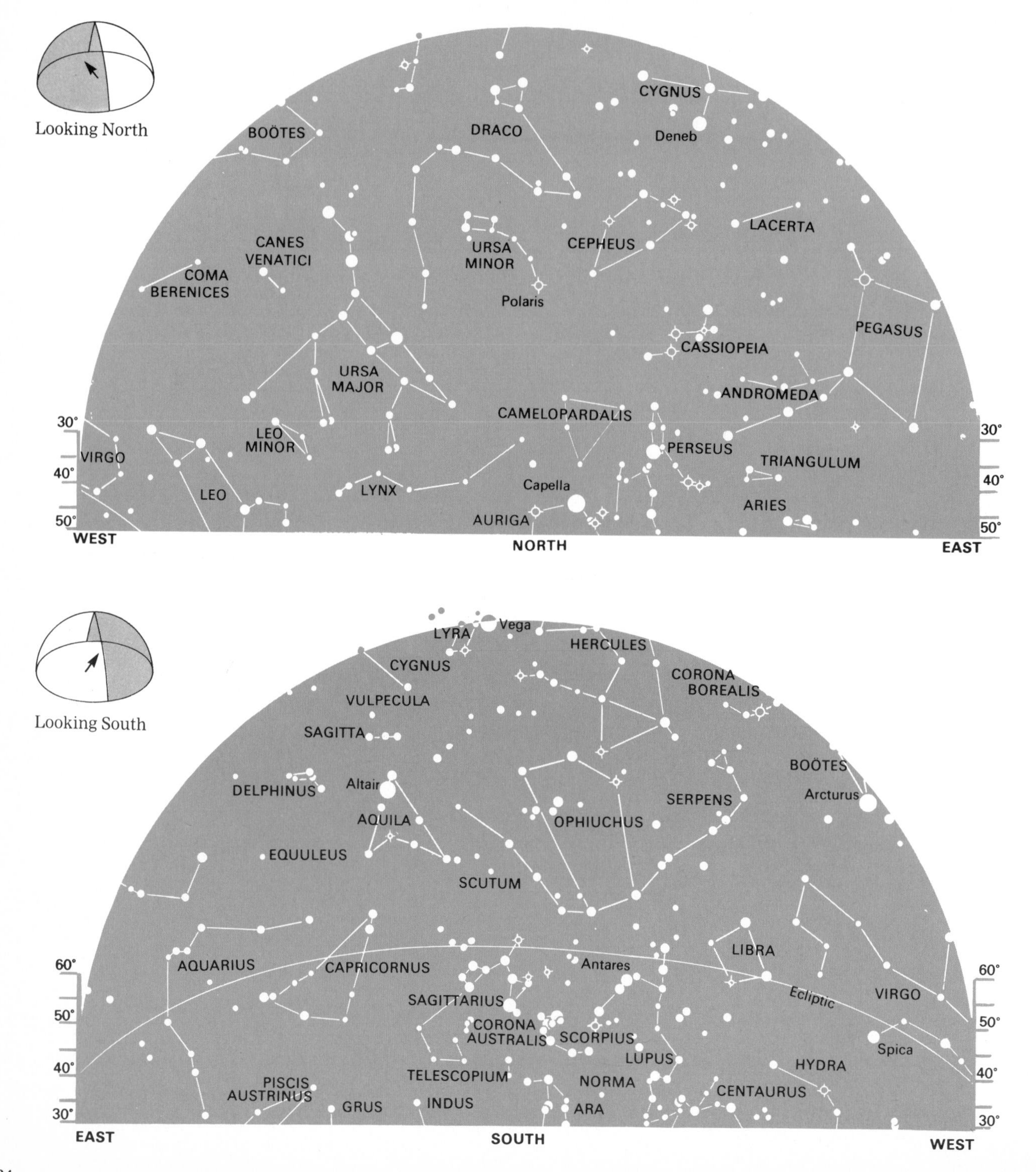

Left: *discovered by Charles Messier in 1764, the Dumbbell is an 8th magnitude planetary nebula. Binoculars and small telescopes will show it as nothing more than a tiny spot, although larger instruments will reveal the conspicuous shape that gave this object its name and may bring out the faint 11th magnitude star situated at the heart of the Dumbbell. This is one of the closest planetary nebulae, its light having taken less than a thousand years to reach us.*

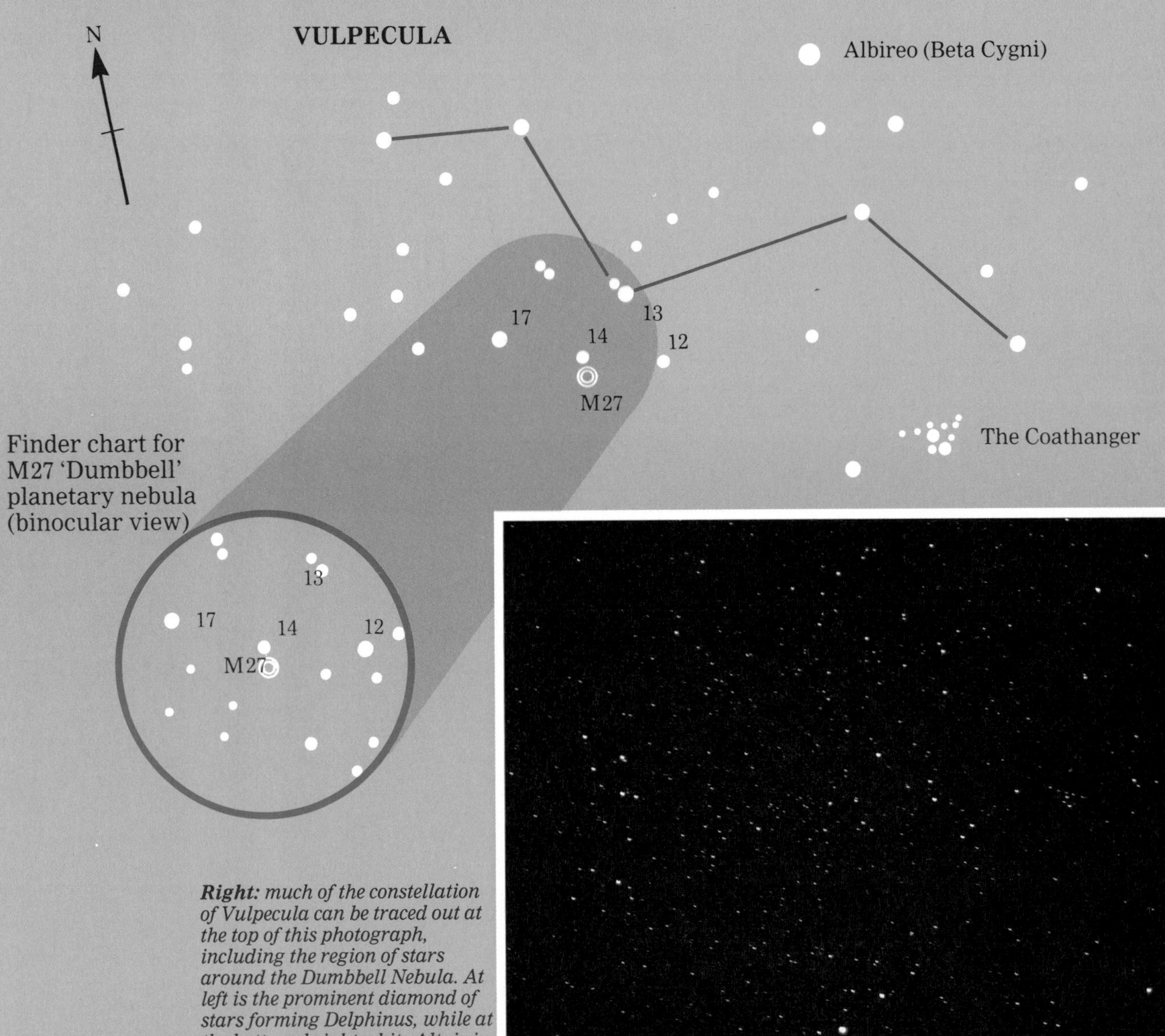

Right: *much of the constellation of Vulpecula can be traced out at the top of this photograph, including the region of stars around the Dumbbell Nebula. At left is the prominent diamond of stars forming Delphinus, while at the bottom, bright white Altair in Aquila is conspicuous. Just right of centre is Sagitta, while the tiny Coathanger cluster is easily visible at the right of the picture.*

***Left:** the conspicuous cruciform shape of Cygnus is well shown in this photograph. The bright star towards the top of the picture is Deneb, a supergiant star around 60,000 times as luminous as our Sun. The light we are seeing from Deneb set off towards us about 1,600 years ago. At the southern end of Cygnus can be found Albireo, one of the most beautiful double stars in the heavens. The golden-yellow primary star has a magnitude of 3.1 and offers a lovely contrast with its fainter companion, which shines at a magnitude of 5.1 with a distinctly blue tint. Their separation is 34.3 seconds of arc, PA 54°. Both components are seen against the back drop of the Milky Way, which passes through Cygnus, and this whole area is well worth sweeping with binoculars.*

***Right:** lying to the west of Cygnus is the small but distinctly shaped constellation of Lyra. Its main star is the brilliant blue-white Vega, the 5th brightest star in the sky. Lying $1\frac{1}{2}°$ northeast of Vega is the famous "Double-double" star Epsilon Lyrae. An observer with keen eyesight will see that Epsilon is a double star, this duplicity being unmistakable with any form of optical aid. This pair has a separation of around 3.5 minutes of arc. Telescopes show that each of these stars is double again, although the two close pairs are not likely to be resolved with telescopes of less than 4 inches aperture. All four stars in the Epsilon system are related to each other.*

***Right:** this photograph shows the region of sky around Deneb together with the open star cluster M39, seen near the top left corner of the picture. When seen through binoculars or a small telescope, M39 has a distinctive "V" shape. It is quite a large, scattered group of stars with around 30 members. It lies at a distance of around 800 light years. M39 can be located easily by following the curved line of stars from Deneb through Rho Cygni. On exceptionally clear nights the cluster may be visible to the naked eye.*

Finder chart for M39 open star cluster (binocular view)

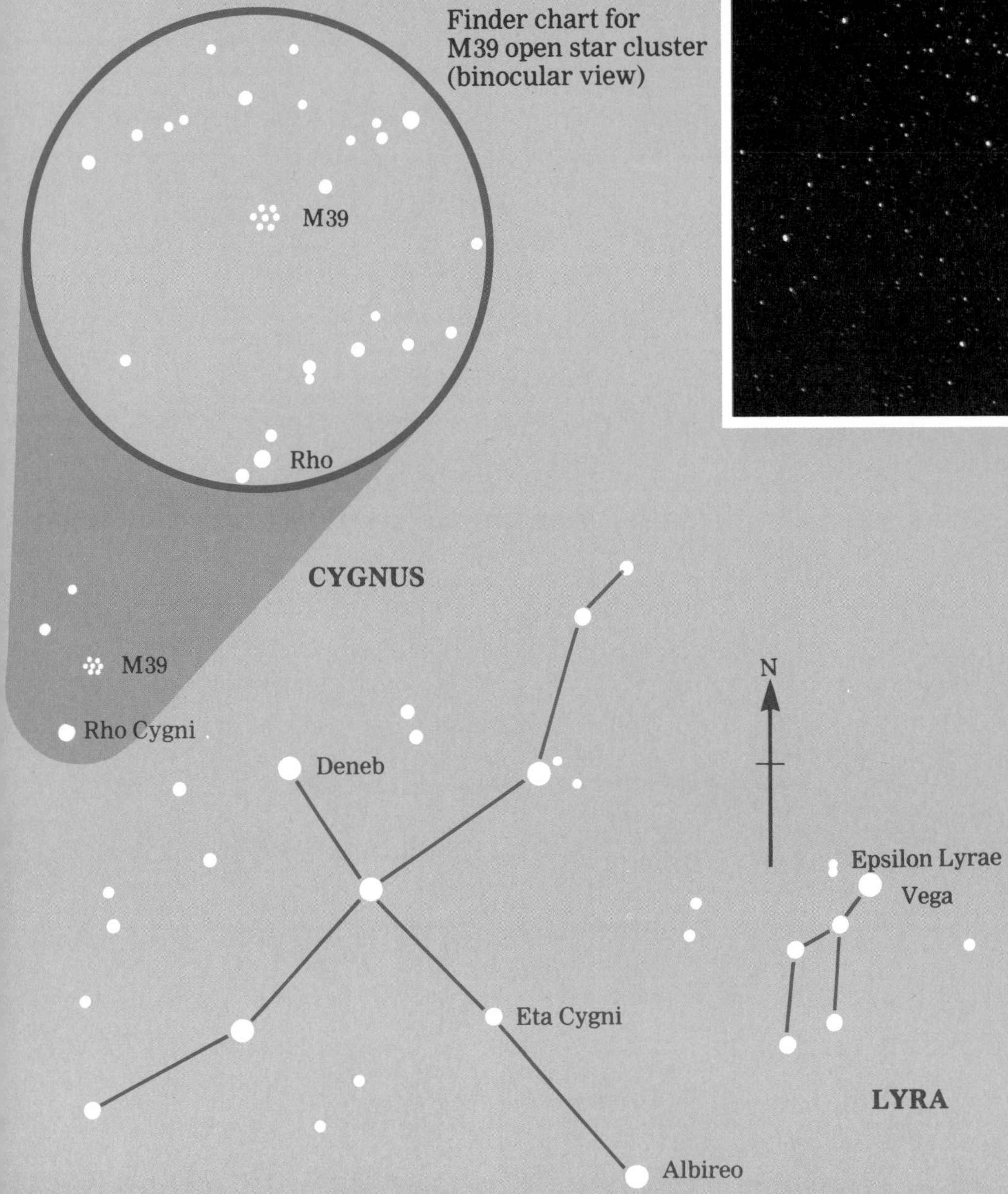

M39
Rho
CYGNUS
M39
Rho Cygni
Deneb
N
Epsilon Lyrae
Vega
Eta Cygni
LYRA
Albireo

The discovery of M39 is often reported to have been made by Le Gentil in 1750, although records indicate that the Greek astronomer Aristotle may have observed it over 2,000 years ago.

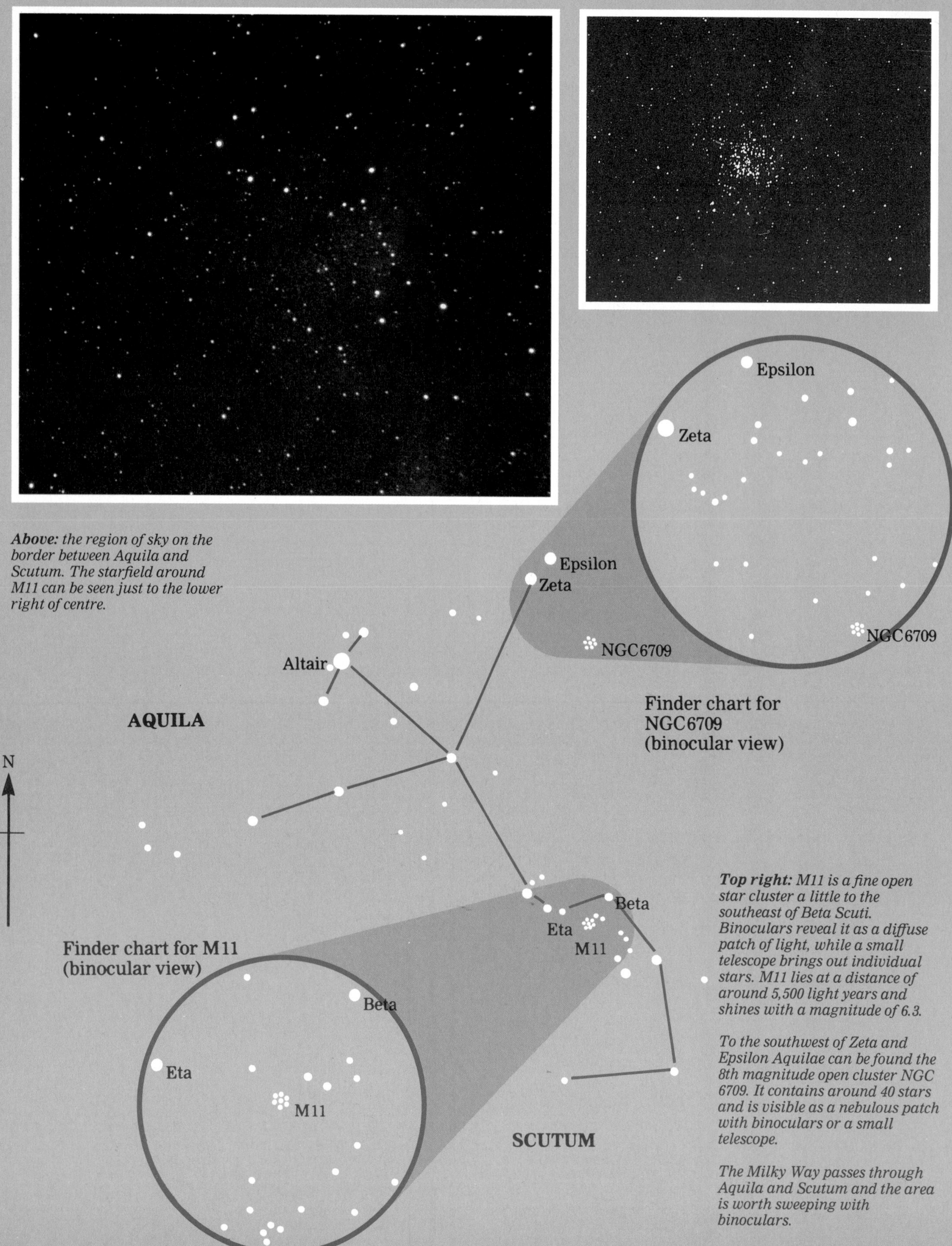

Above: *the region of sky on the border between Aquila and Scutum. The starfield around M11 can be seen just to the lower right of centre.*

Top right: *M11 is a fine open star cluster a little to the southeast of Beta Scuti. Binoculars reveal it as a diffuse patch of light, while a small telescope brings out individual stars. M11 lies at a distance of around 5,500 light years and shines with a magnitude of 6.3.*

To the southwest of Zeta and Epsilon Aquilae can be found the 8th magnitude open cluster NGC 6709. It contains around 40 stars and is visible as a nebulous patch with binoculars or a small telescope.

The Milky Way passes through Aquila and Scutum and the area is worth sweeping with binoculars.

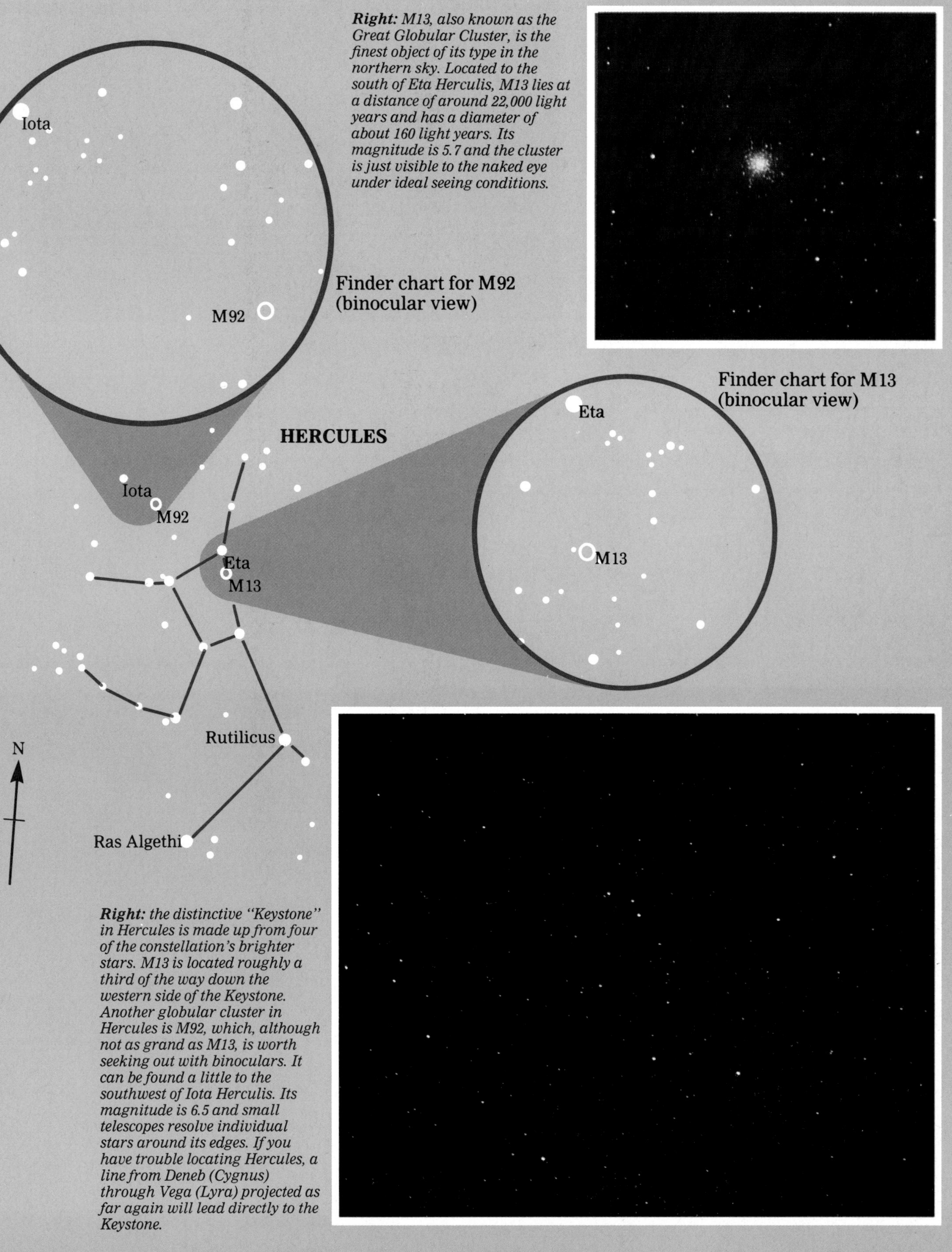

Right: *M13, also known as the Great Globular Cluster, is the finest object of its type in the northern sky. Located to the south of Eta Herculis, M13 lies at a distance of around 22,000 light years and has a diameter of about 160 light years. Its magnitude is 5.7 and the cluster is just visible to the naked eye under ideal seeing conditions.*

Right: *the distinctive "Keystone" in Hercules is made up from four of the constellation's brighter stars. M13 is located roughly a third of the way down the western side of the Keystone. Another globular cluster in Hercules is M92, which, although not as grand as M13, is worth seeking out with binoculars. It can be found a little to the southwest of Iota Herculis. Its magnitude is 6.5 and small telescopes resolve individual stars around its edges. If you have trouble locating Hercules, a line from Deneb (Cygnus) through Vega (Lyra) projected as far again will lead directly to the Keystone.*

***Right:** the Trifid Nebula (M20) is a somewhat elusive object for the binocular observer. Its magnitude is 6.9 and it can be found to the southwest of Mu Sagittarii. The Trifid Nebula derives its name from the distinctive lanes of dust that seem to divide the gas cloud into three. The distance of the Trifid is not known with certainty, although recent estimates put it at around 5,000 light years. Less than a degree to the northeast of M20 is the open star cluster M21, not shown on this photograph. The Lagoon Nebula (M8) is also nearby, situated a few degrees to the southeast.*

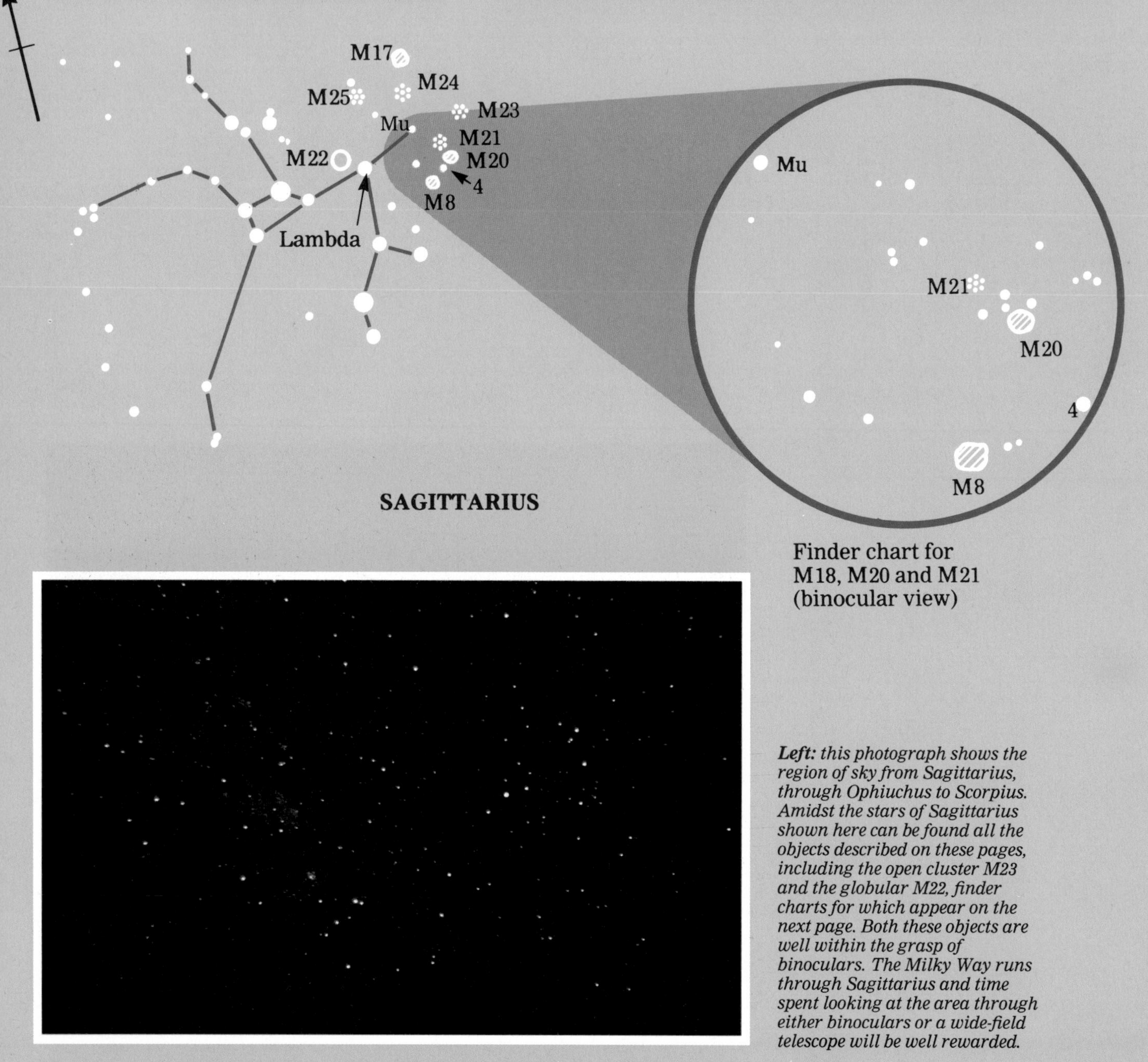

Finder chart for M18, M20 and M21 (binocular view)

***Left:** this photograph shows the region of sky from Sagittarius, through Ophiuchus to Scorpius. Amidst the stars of Sagittarius shown here can be found all the objects described on these pages, including the open cluster M23 and the globular M22, finder charts for which appear on the next page. Both these objects are well within the grasp of binoculars. The Milky Way runs through Sagittarius and time spent looking at the area through either binoculars or a wide-field telescope will be well rewarded.*

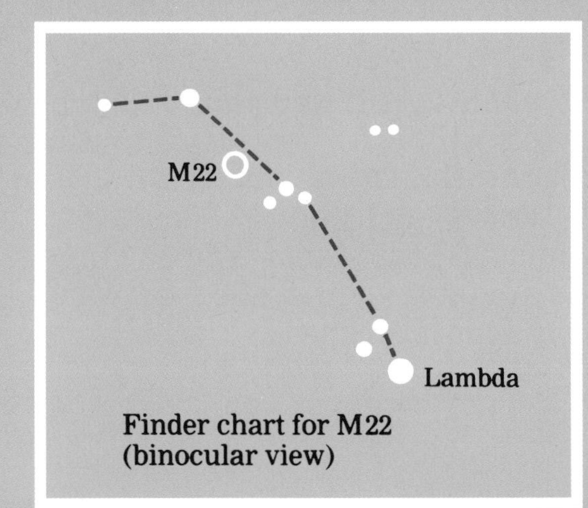

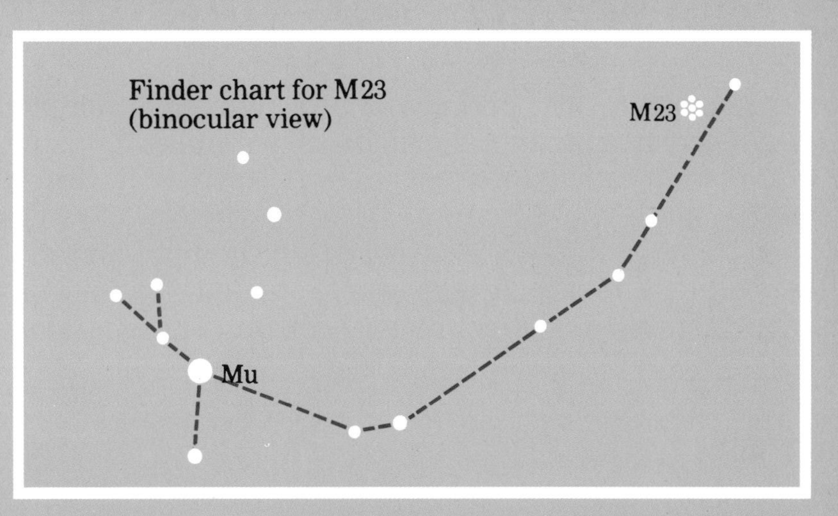

Right: *starfield around the Swan or Omega Nebula (M17). Lying at a distance of almost 6,000 light years, M17 has a magnitude of 7.5 and can be seen a few degrees to the north of Mu Sagittarii.*
Below: *between Mu and M17 are the open star cluster M18 and the Milky Way star cloud M24. M18 has a magnitude of 7.5 and lies at a distance of almost 5,000 light years. Discovered by Charles Messier in 1764 it contains about a dozen stars. M24 isn't a true cluster but only a bright star cloud. M25 is located to the northeast of Mu. This is another open star cluster containing at least 50 stars.*

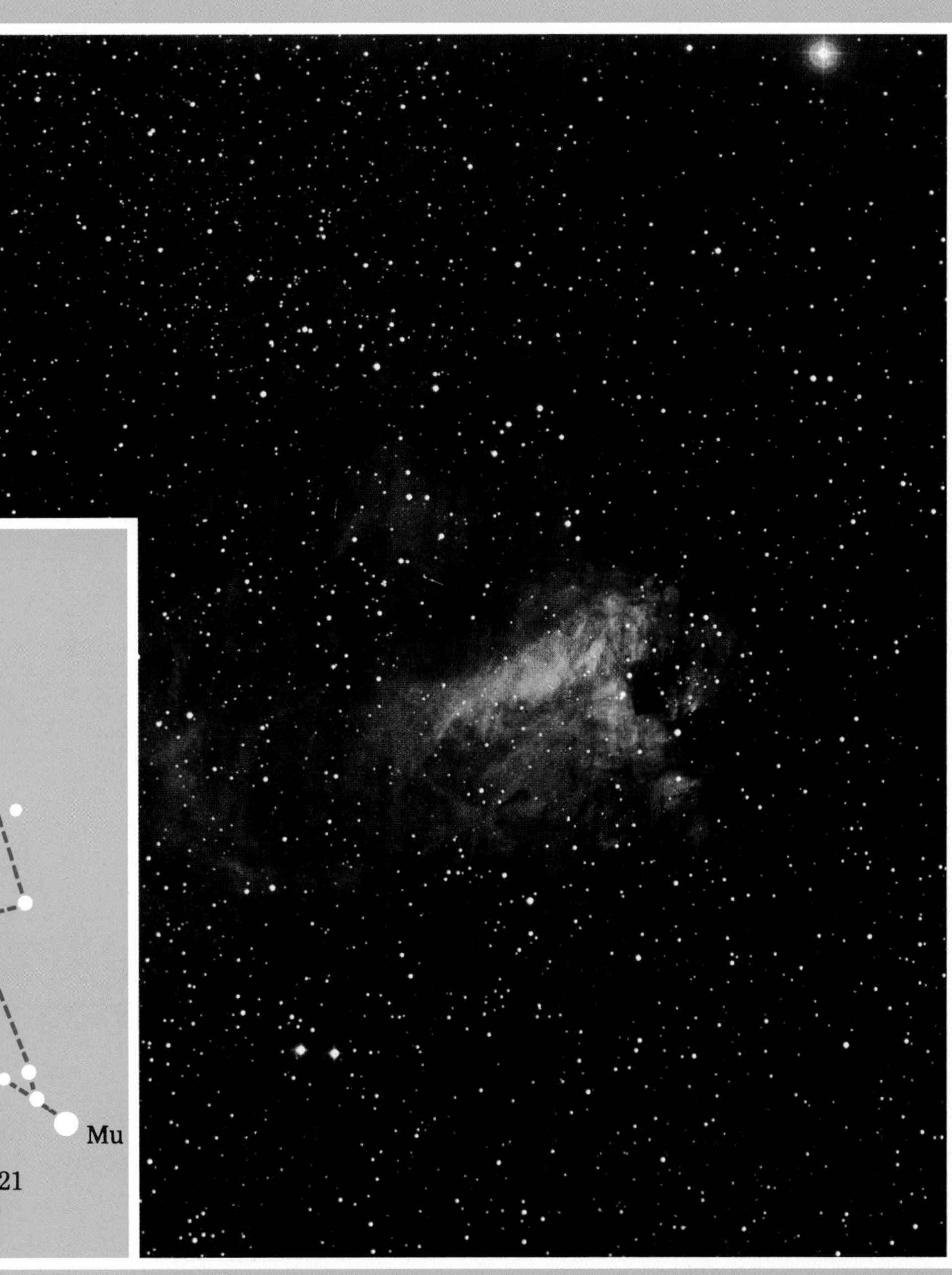

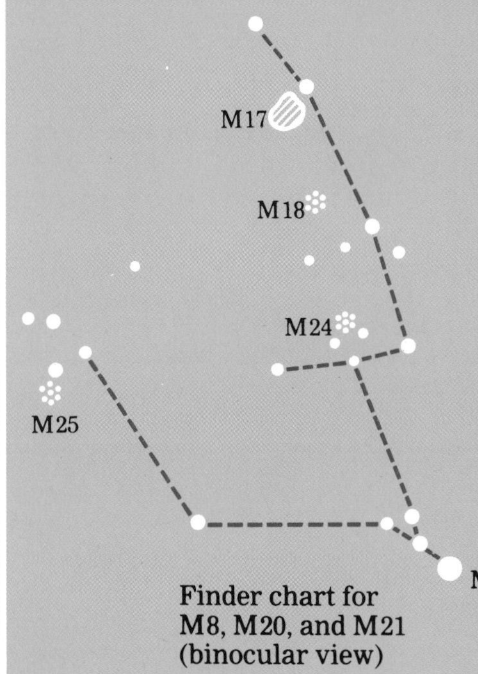

The Northern Sky: Autumn

Autumn skies are perhaps a little barren, situated between the starfields of summer and the majestic winter constellations. High in the south the Square of Pegasus can be seen, the two stars on its western side acting as pointers to the bright star Fomalhaut, seen just above the southern horizon. The Summer Triangle is still well placed high in the southwest, although Ursa Major is well down in the north. To the northeast Aldebaran in Taurus and Capella in Auriga herald winter. Above them Perseus and Andromeda are prominent, while the "W" of Cassiopeia is almost overhead.

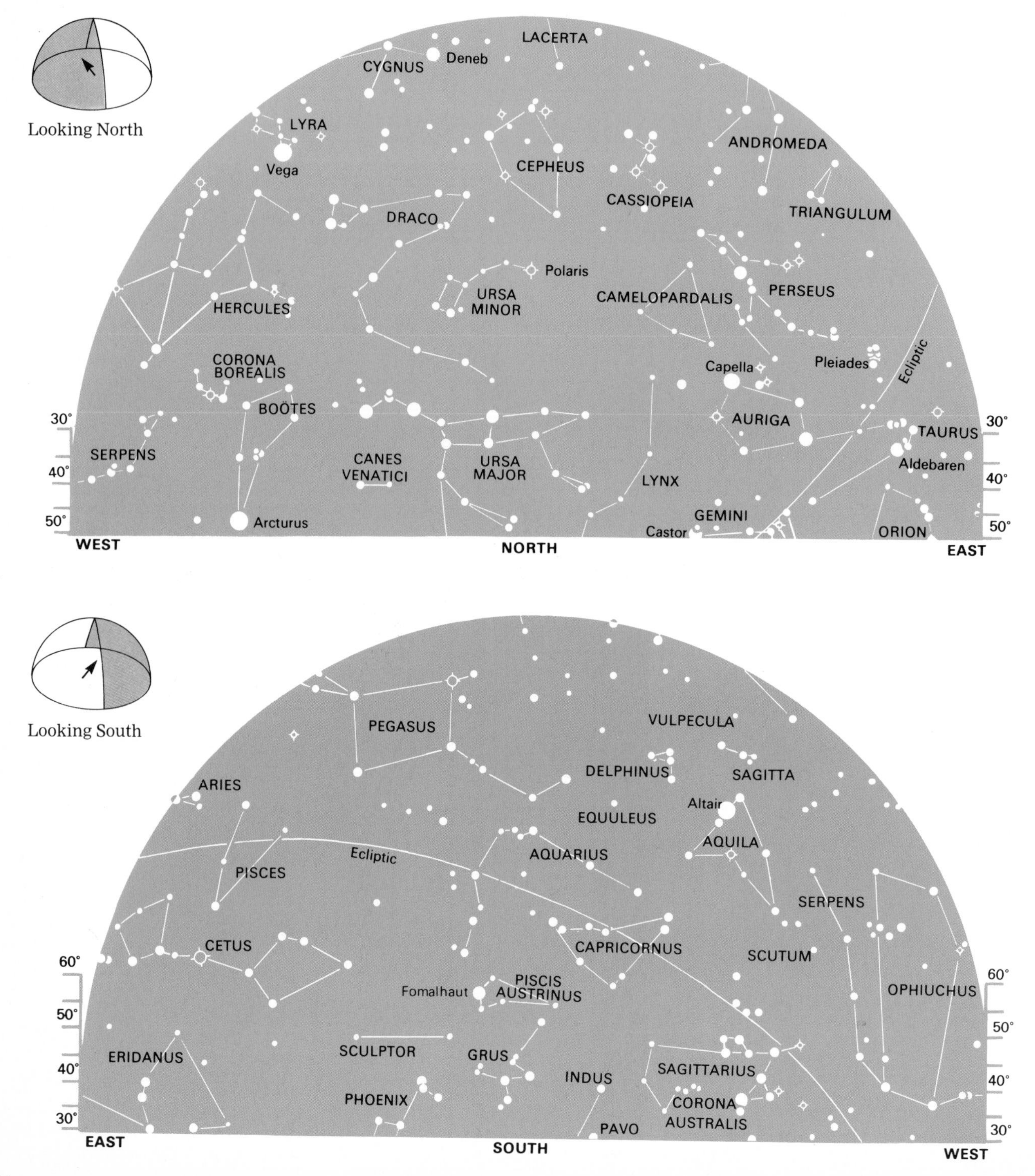

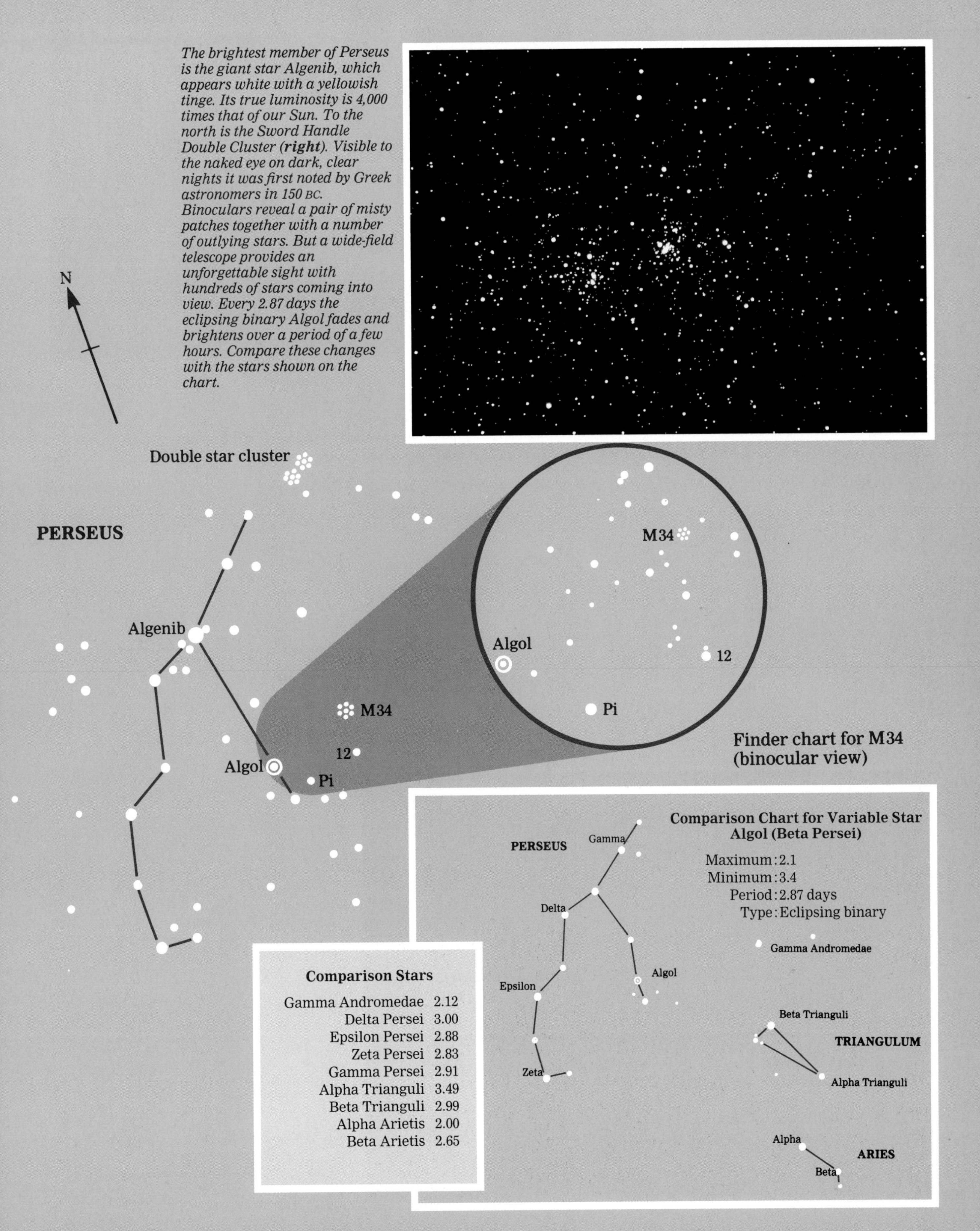

*The brightest member of Perseus is the giant star Algenib, which appears white with a yellowish tinge. Its true luminosity is 4,000 times that of our Sun. To the north is the Sword Handle Double Cluster (**right**). Visible to the naked eye on dark, clear nights it was first noted by Greek astronomers in 150 BC. Binoculars reveal a pair of misty patches together with a number of outlying stars. But a wide-field telescope provides an unforgettable sight with hundreds of stars coming into view. Every 2.87 days the eclipsing binary Algol fades and brightens over a period of a few hours. Compare these changes with the stars shown on the chart.*

Comparison Stars

Star	Magnitude
Gamma Andromedae	2.12
Delta Persei	3.00
Epsilon Persei	2.88
Zeta Persei	2.83
Gamma Persei	2.91
Alpha Trianguli	3.49
Beta Trianguli	2.99
Alpha Arietis	2.00
Beta Arietis	2.65

Right: *situated just to the west of Nu Andromedae, the Andromeda Galaxy (M31) is the brightest spiral galaxy in the sky. It is also the closest, lying at a distance of only 2.2 million light years. M31 is a naked-eye object and records of sightings go back to the 10th century, although the astronomers of that time had no idea as to the true identity of the tiny cloud they saw. Current research shows that M31 contains around 300,000 million stars and has a diameter in the region of 150,000 light years.*

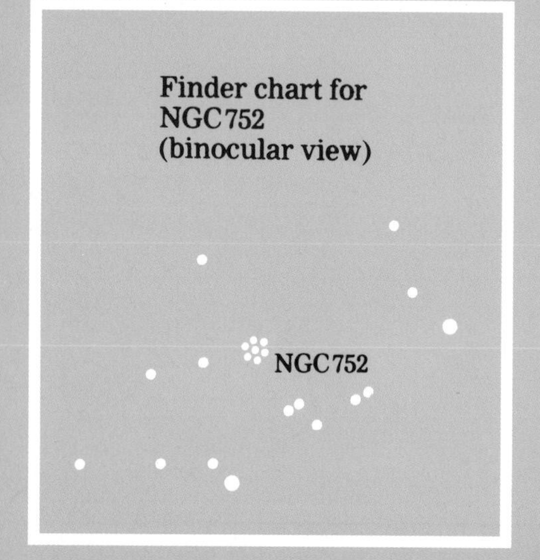

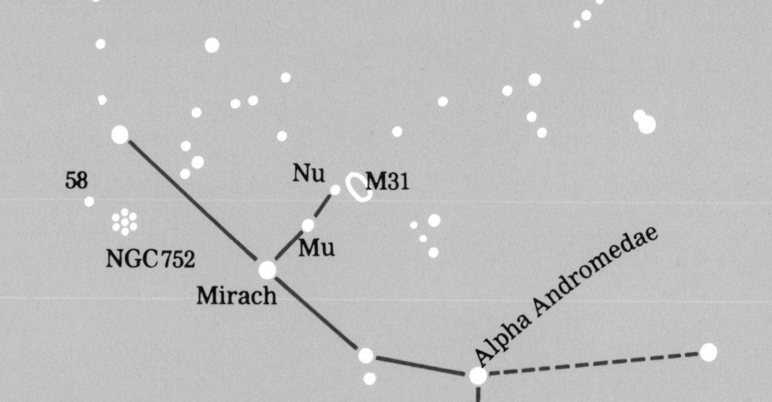

Left: *the region of sky around M31. The stars Mirach, Mu and Nu can be seen near centre with M31 visible as a small elongated patch. Binoculars show the nucleus together with an indication of spiral arms. However, it is something of a disappointment as we see M31 almost edge-on, and the full beauty of this gigantic island universe is lost. Accompanying M31 are two small elliptical galaxies. M32 lies a little to the south, while M110 can be seen slightly further away to the north. Both are faint and telescopes are required to see them.*
To the west of 58 Andromedae is the open cluster NGC 752. It is quite large, containing a number of bright stars, and is an easy binocular object.

Right: *Pegasus is a fairly prominent constellation, some of its brighter stars, together with Alpha Andromedae, forming a large square that is visible high in the south during autumn evenings. However, prominent as it may be, it contains little to interest the observer. Pi Pegasi is a wide binocular double with white and yellow components of magnitudes 4.4 and 5.7, the fainter star being to the west of its brighter companion. Enif is a giant star of magnitude 2.3, which has a distinct yellow-orange tint. Its true luminosity is around 6,000 times that of the Sun.*

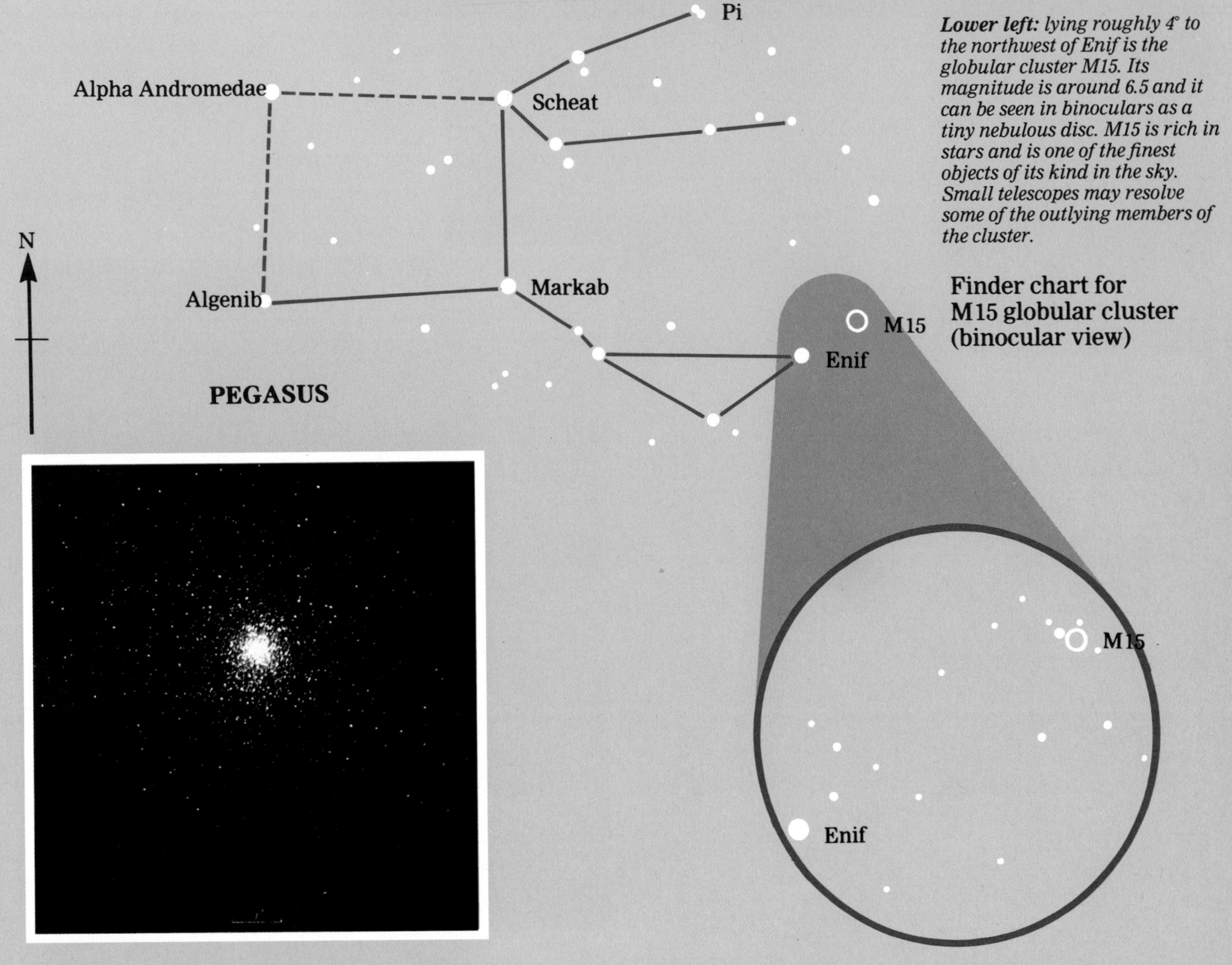

Lower left: *lying roughly 4° to the northwest of Enif is the globular cluster M15. Its magnitude is around 6.5 and it can be seen in binoculars as a tiny nebulous disc. M15 is rich in stars and is one of the finest objects of its kind in the sky. Small telescopes may resolve some of the outlying members of the cluster.*

Left: *this photograph shows the area of sky around Aries and Triangulum. Gamma Arietis is a fine telescopic double with components of magnitudes 4.75 and 4.83. They are both white and lie 7.8 seconds of arc apart, PA 360°. These two stars lie at a distance of around 160 light years and have a combined true luminosity of about 50 times that of our Sun.*
Below: *the Triangulum Spiral (M33) is a large face-on galaxy situated at a distance of 2.4 million light years. It is difficult to see unless the sky is really dark and clear, when it can be seen with the naked eye under ideal conditions. Most will need some form of optical aid to see M33, which will appear as a dim and hazy patch of light.*

M33

Metallah

Finder chart for M33 spiral galaxy (binocular view)

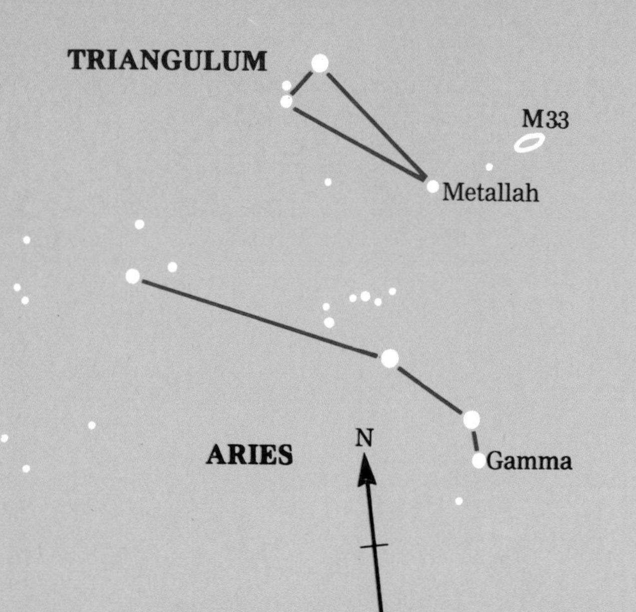

__Right:__ Mira (Omicron Ceti) is the brightest and best known long-period variable. As a general rule it reaches 3rd magnitude at maximum, although on rare occasions it has reached 2nd magnitude. The chart shows suitable stars with which to compare the brightness of Mira as it fades and brightens. Only moderate optical aid is required.

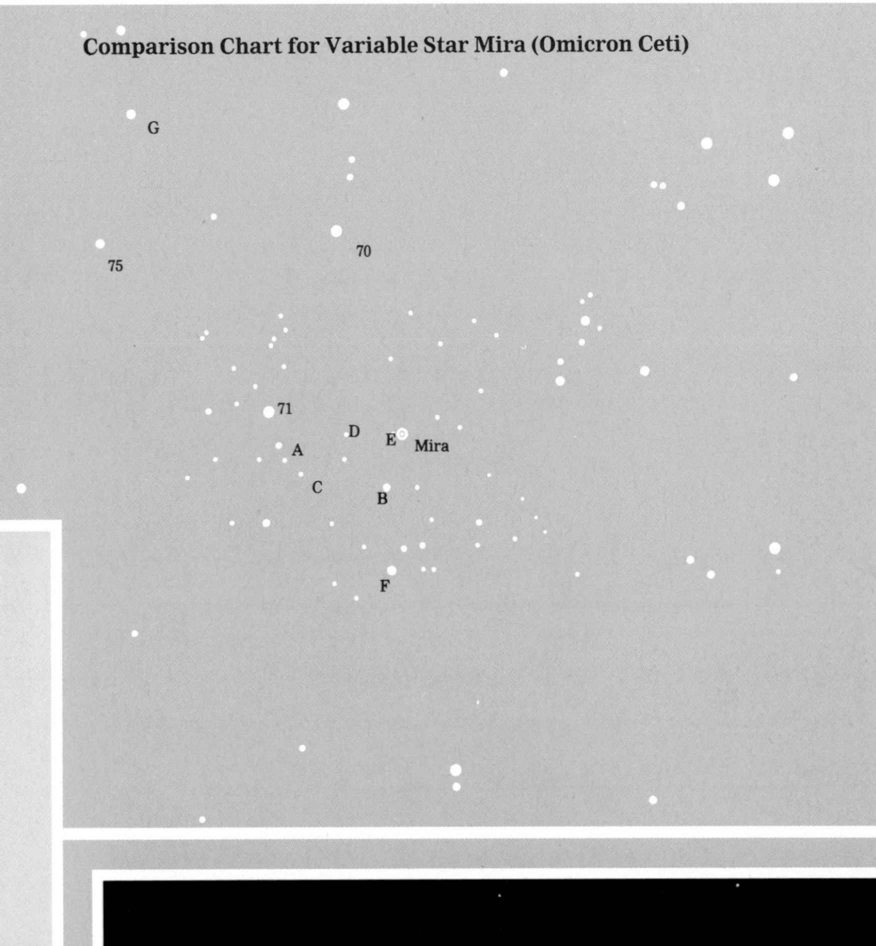

MIRA (OMICRON CETI)

Maximum: 2nd or 3rd magnitude
Minimum: 9th magnitude
Period: 331 days average
Type: Long-period pulsating variable

Comparison Stars

Beta Ceti	2.00	75 Ceti	5.34
Alpha Ceti	2.52	70 Ceti	5.41
Eta Ceti	3.44	G	6.00
Mu Ceti	4.26	71 Ceti	6.40
Xi$_2$ Ceti	4.27	F	6.49
Xi$_1$ Ceti	4.36	A	7.19
Lambda Ceti	4.70	B	8.00
Nu Ceti	4.87	C	8.60
		D	8.80
		E	9.20

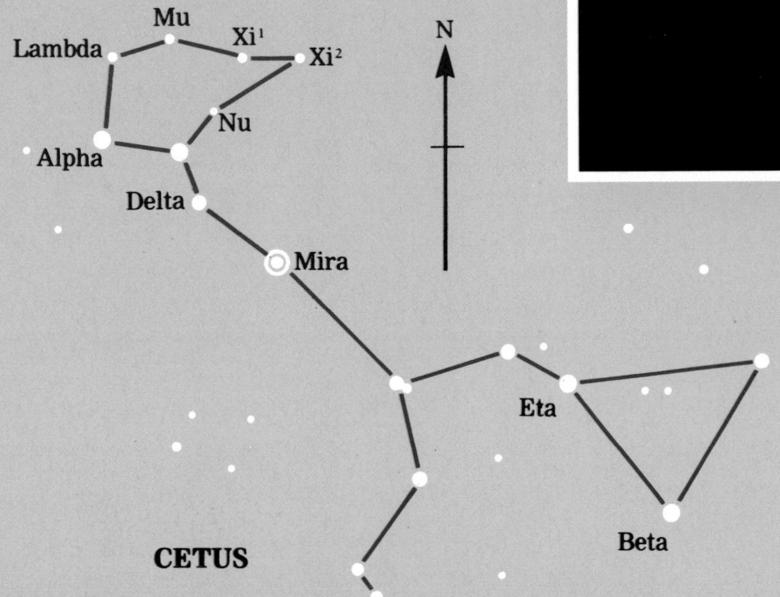

__Above:__ although Cetus is quite a large constellation it has few objects of interest within its boundaries. Alpha Ceti has a conspicuous orange-red tint. This contrasts with the blue-white Delta and the yellow-orange hue of Eta. The latter is part of a double system with a 9th magnitude companion situated 3.5 minutes of arc away. By far the most interesting object in Cetus is the long-period variable star Mira, details of which are given above.

***Below:** Cassiopeia contains quite a few open star clusters, including NGC 457 and NGC 663. NGC 457 can be seen next to the star 34 Cassiopeiae. It contains around 100 stars and shines with a magnitude of 7.5 from a distance of more than 9,000 light years. It will be seen as a diffuse patch of light in binoculars, a telescope being required in order to resolve individual stars. NGC 663 is a little brighter at magnitude 7.1. This cluster contains somewhere in the region of 80 stars and lies at a distance of about 2,600 light years.*

***Above:** from an observational point of view, M103 is a rather poor example of an open star cluster. It contains around 60 stars and shines with a magnitude of 7.4. Seen quite close to the star Ksora, M103 is situated at a distance of over 8,000 light years. Telescopically, it takes on the appearance of a wedge or fan-shaped group of stars. All the clusters described on this page can be found quite easily by using the accompanying charts.*

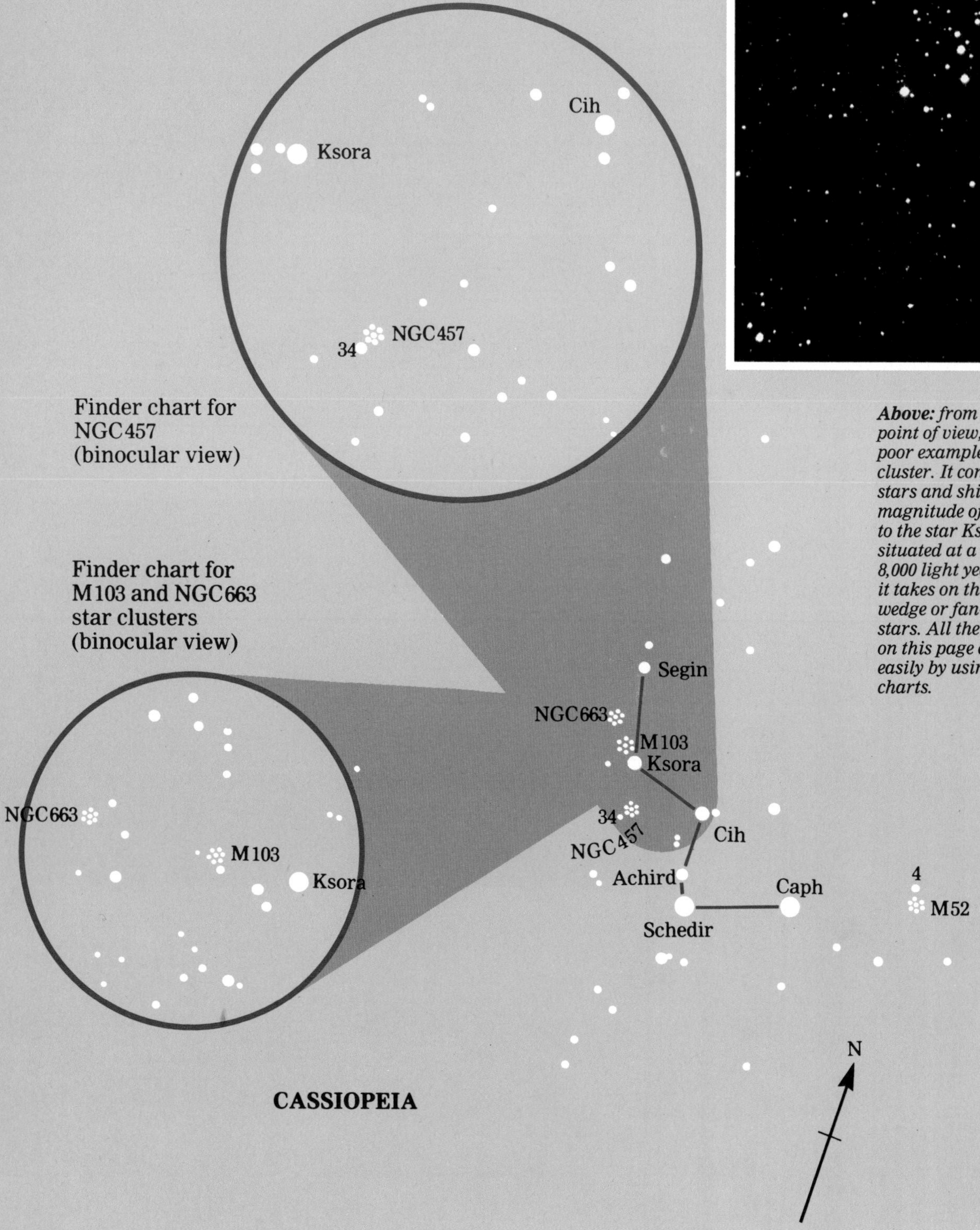

Finder chart for NGC457 (binocular view)

Finder chart for M103 and NGC663 star clusters (binocular view)

CASSIOPEIA

***Right:** Cassiopeia lies within the Milky Way and contains many beautiful starfields. It takes the form of a prominent "W", the brightest member of which is Schedir, an orange star whose colour is easily seen in binoculars. Schedir is slightly variable, although the small changes in brightness are somewhat difficult to follow. Its distance is estimated to be almost 200 light years and its true luminosity to be 230 times that of our Sun. Achird is a beautiful double star, both components of which can be seen through a small telescope. These two stars actually form a binary system and orbit each other once every 500 years or so. Their current separation is around 11 seconds of arc. The most recent closest approach of the two stars to each other was in 1889, and since then they have been slowly moving apart and becoming easier to resolve.*

***Below:** M52 is a magnificent open star cluster. It was discovered by Charles Messier in 1774 and contains around 200 stars. M52 shines with a magnitude of 7.3, its light having taken over 5,000 years to reach us. It is an easy object for binoculars or a small telescope.*

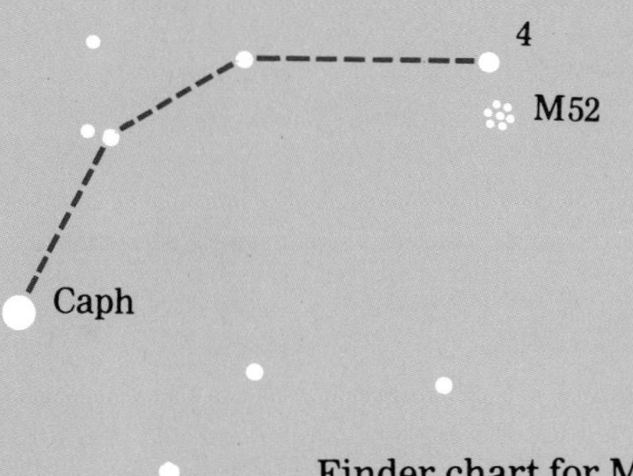

Finder chart for M52 (binocular view)

The Southern Sky: Winter

Most of the northern aspect is taken up by the three large but obscure groups Hercules, Ophiuchus and Serpens, although the small circlet of stars forming Corona Borealis, the Northern Crown, is prominent close to Hercules. Its southern counterpart, Corona Australis, is visible almost overhead. The distinctive shape of Scorpius and his sting is found next to the teeming star fields of Sagittarius. The two bright stars Achernar (Eridanus) and Fomalhaut (Pisces Austrinus) are well displayed in the southeastern sky.

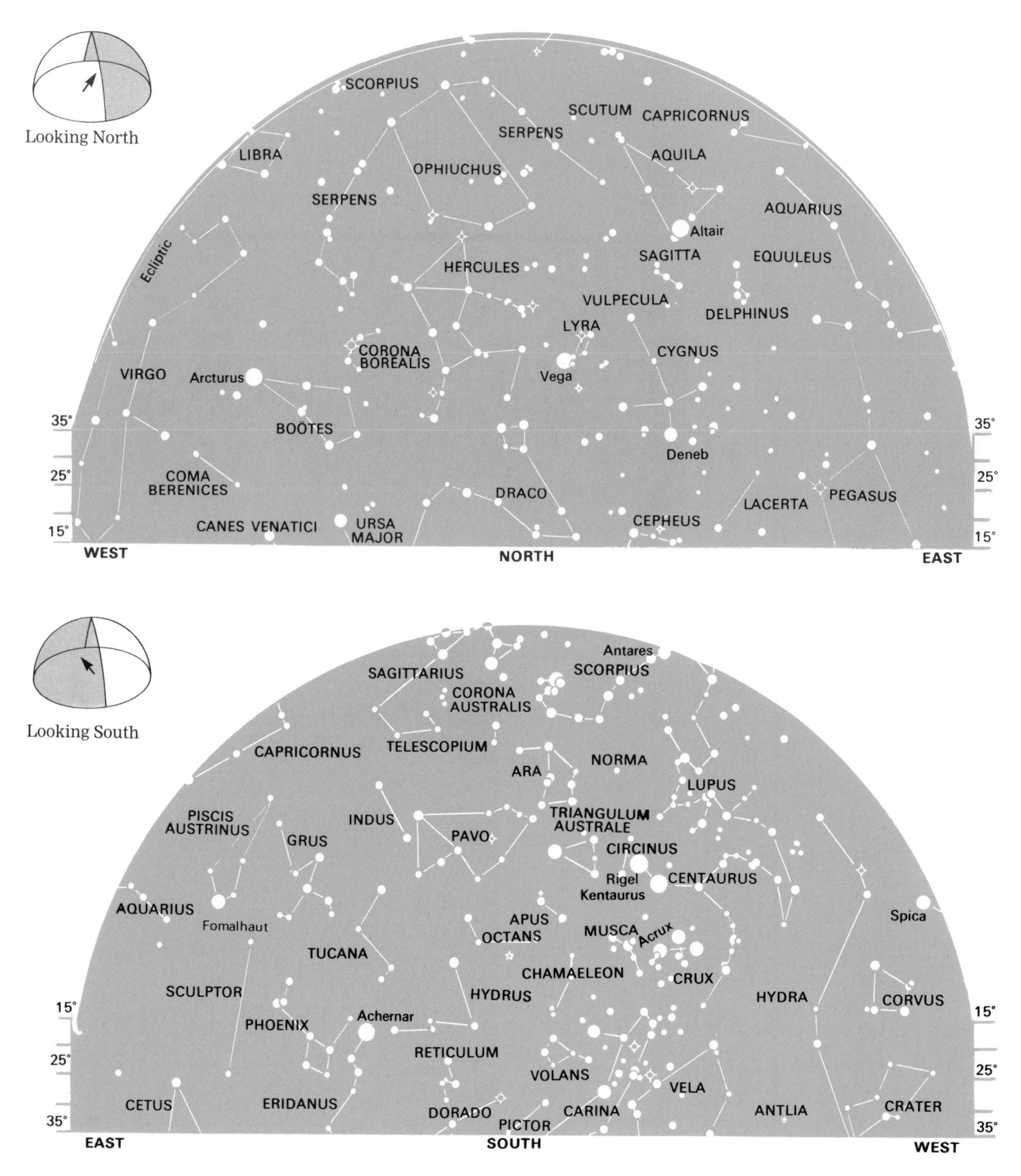

NGC 6752, located close to Omega Pavonis, is an excellent example of a globular cluster. Discovered in 1828 its magnitude of 7.2 makes it an easy object for binocular observers. NGC 6752 lies at a distance of around 20,000 light years, making it one of the closest globulars in the sky.

Kappa Pavonis is an interesting variable star of the Cepheid type. Useful comparison stars can be found quite nearby and the variations in brightness of Kappa can be followed either with binoculars or the naked eye.

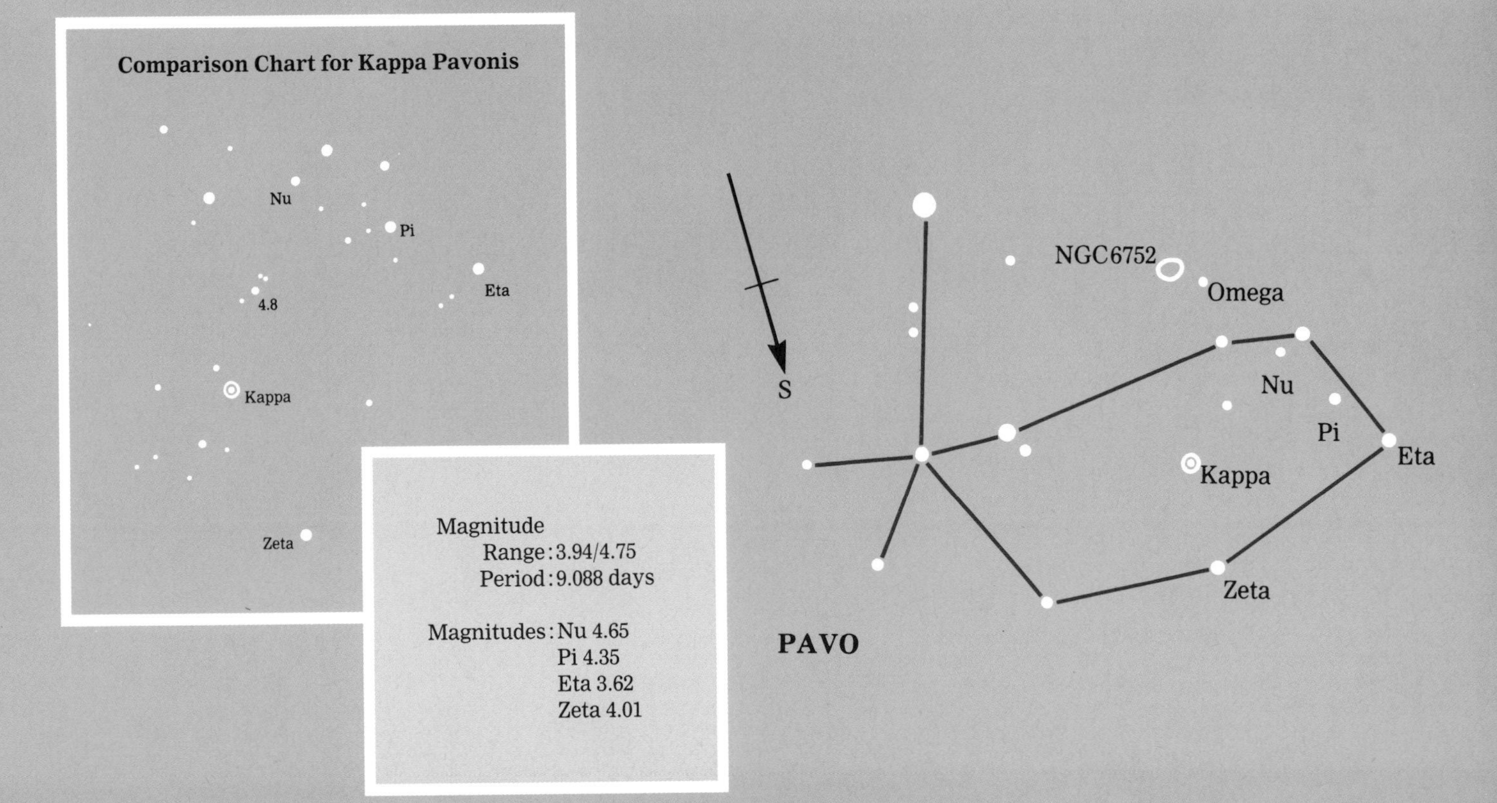

***Right:** the constellation Ara has quite a distinctive shape and contains a number of clusters that lie within the light grasp of binoculars. One of the best is NGC 6193, a 5th magnitude open cluster which contains around 30 stars. Zeta Arae shines with a magnitude of 3.16 and has a distinctly orange hue. It lies at a distance of around 90 light years and has a true luminosity of about 35 times that of our Sun.*

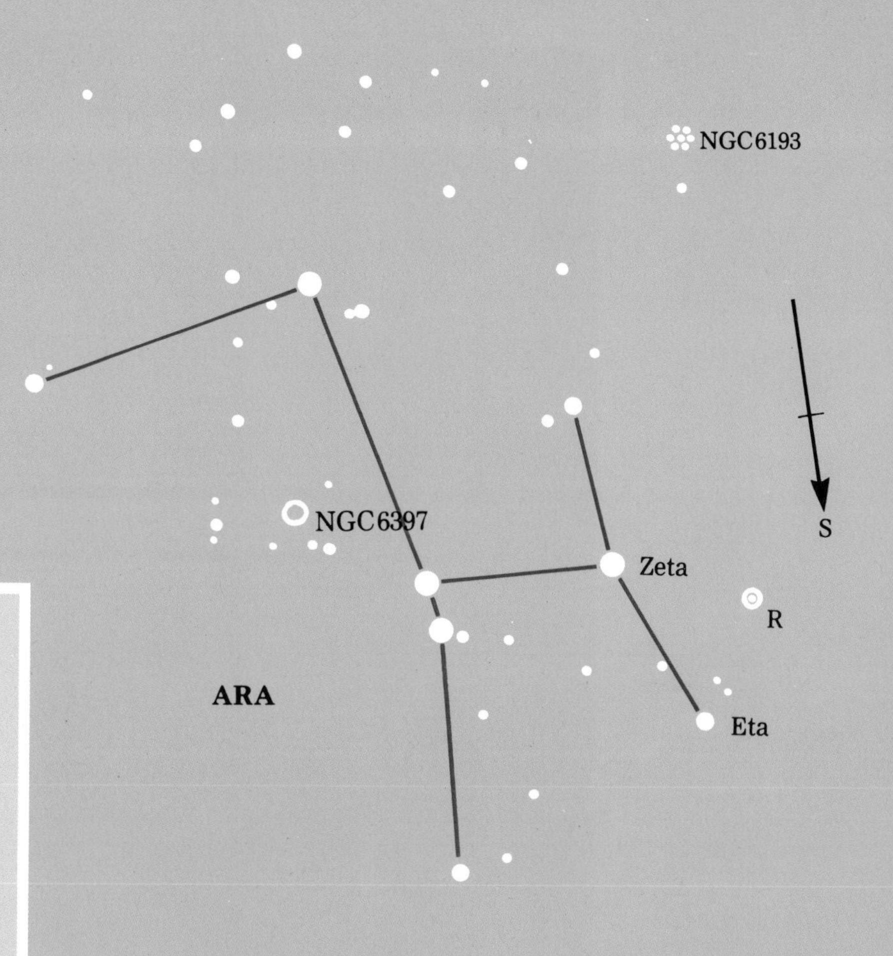

R ARAE

Maximum: 6.0
Minimum: 7.0
Period: 4.425 days
Type: Algol

Comparison Stars

A	6.5
B	7.0
C	7.0
D	6.0
E	7.3
F	6.9
G	6.6

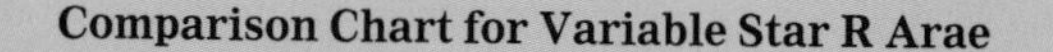

Comparison Chart for Variable Star R Arae

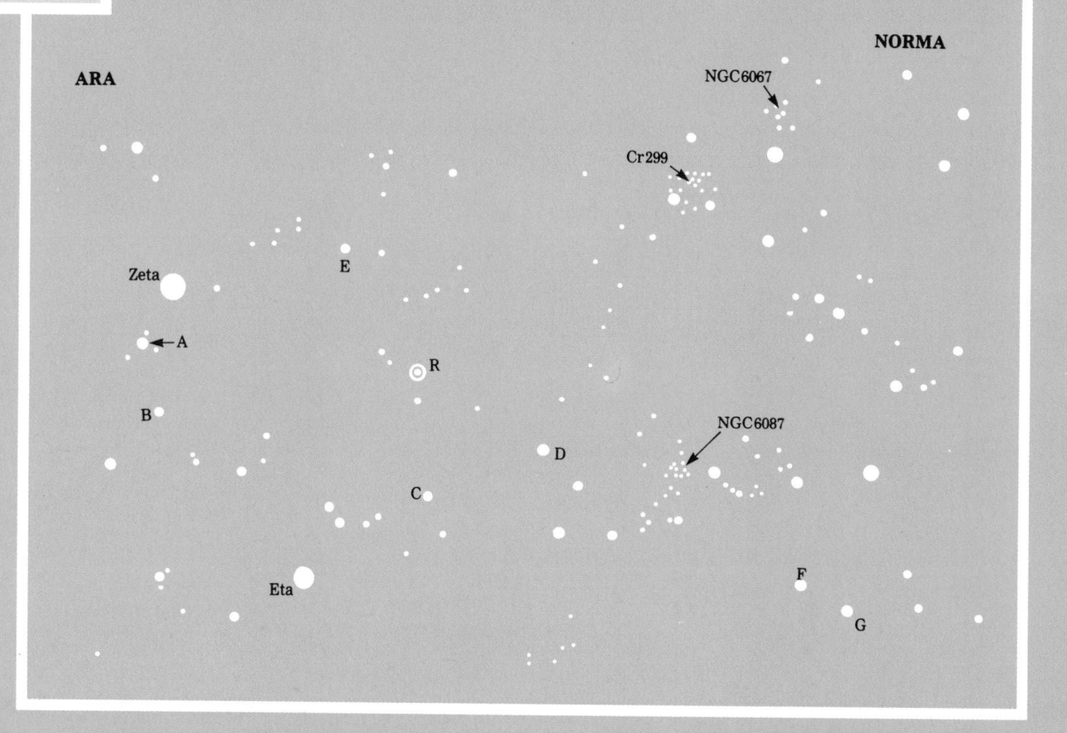

***Right:** R Arae is an eclipsing binary which lies in the same binocular field as Zeta and Eta Arae. The accompanying chart shows a number of suitable comparison stars together with part of the nearby constellation of Norma, which plays host to quite a few star clusters within the light grasp of binoculars. NGC 6087 is rich in stars, and at magnitude 6.0 is marginally brighter than NGC 6067 seen further to the north. NGC 6067 contains over 100 stars and shines from a distance of nearly 7,000 light years with a magnitude of 6.7. Although poor in stars Cr299, which lies in the same area, is worth seeking out.*

Above: *this photograph shows the constellation of Ara and includes the field around NGC 6397, a bright globular cluster which was discovered by Lacaille in 1755. Its distance is estimated to be 8,200 light years, making it the closest object of its kind in the sky. Ara and Norma lie across the Milky Way and the area is well worth sweeping either with binoculars or a wide-field telescope.*

The Southern Sky: Spring

High in the north the Square of Pegasus points the way to Fomalhaut, the brightest star in Pisces Austrinus. Almost overhead Grus the Crane, Tucana the Toucan, Pavo the Peacock and the legendary Phoenix wheel around the sky. The Phoenix is perched on the banks of the River Eridanus. The brilliant white Canopus, leading member of Carina and second brightest star in the sky, can be seen low over the southern horizon. In the north Vega in Lyra, Deneb in Cygnus and Altair in Aquila form a large triangle through which passes the northern Milky Way.

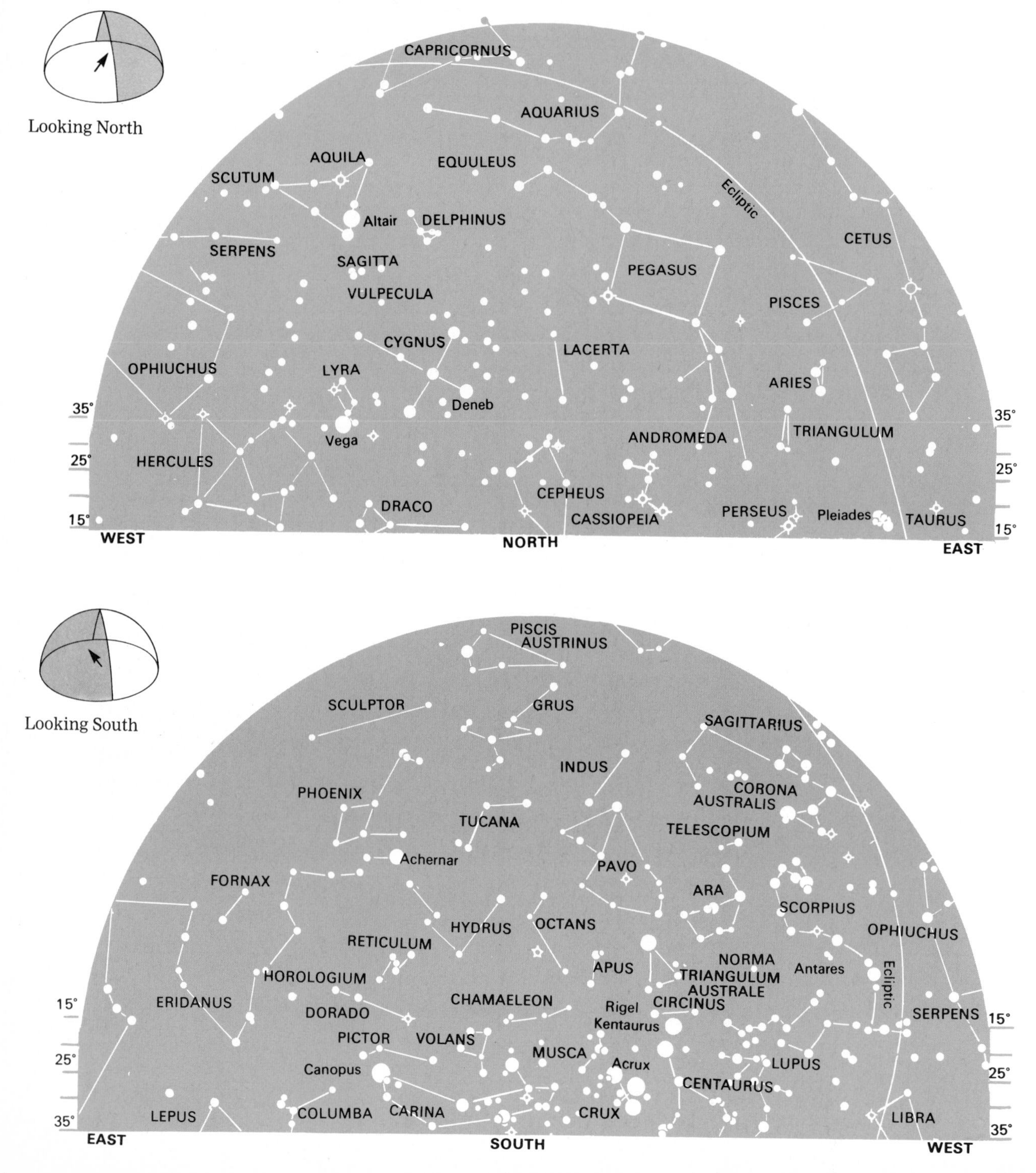

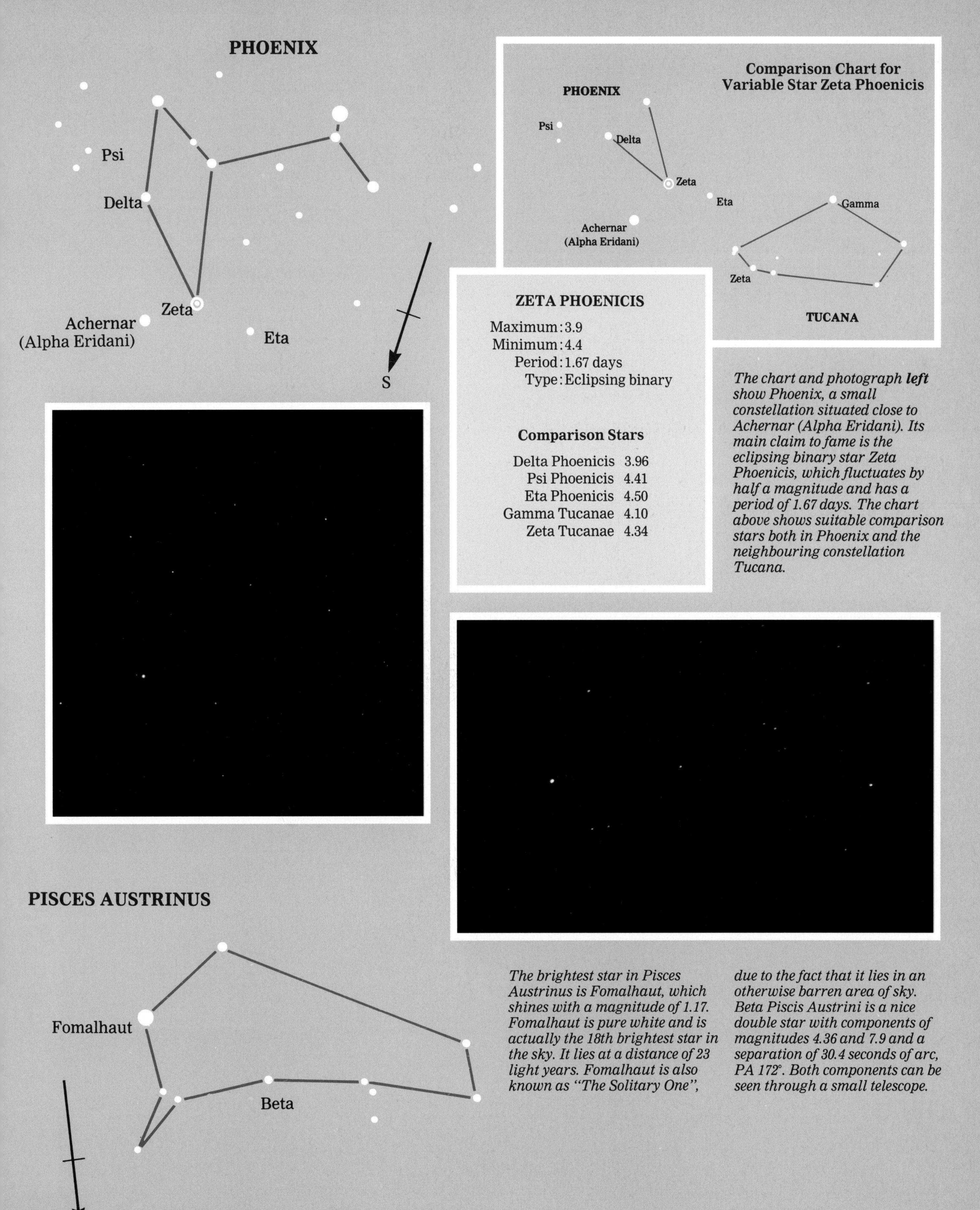

ZETA PHOENICIS

Maximum:	3.9
Minimum:	4.4
Period:	1.67 days
Type:	Eclipsing binary

Comparison Stars

Delta Phoenicis	3.96
Psi Phoenicis	4.41
Eta Phoenicis	4.50
Gamma Tucanae	4.10
Zeta Tucanae	4.34

*The chart and photograph **left** show Phoenix, a small constellation situated close to Achernar (Alpha Eridani). Its main claim to fame is the eclipsing binary star Zeta Phoenicis, which fluctuates by half a magnitude and has a period of 1.67 days. The chart above shows suitable comparison stars both in Phoenix and the neighbouring constellation Tucana.*

The brightest star in Pisces Austrinus is Fomalhaut, which shines with a magnitude of 1.17. Fomalhaut is pure white and is actually the 18th brightest star in the sky. It lies at a distance of 23 light years. Fomalhaut is also known as "The Solitary One", due to the fact that it lies in an otherwise barren area of sky. Beta Piscis Austrini is a nice double star with components of magnitudes 4.36 and 7.9 and a separation of 30.4 seconds of arc, PA 172°. Both components can be seen through a small telescope.

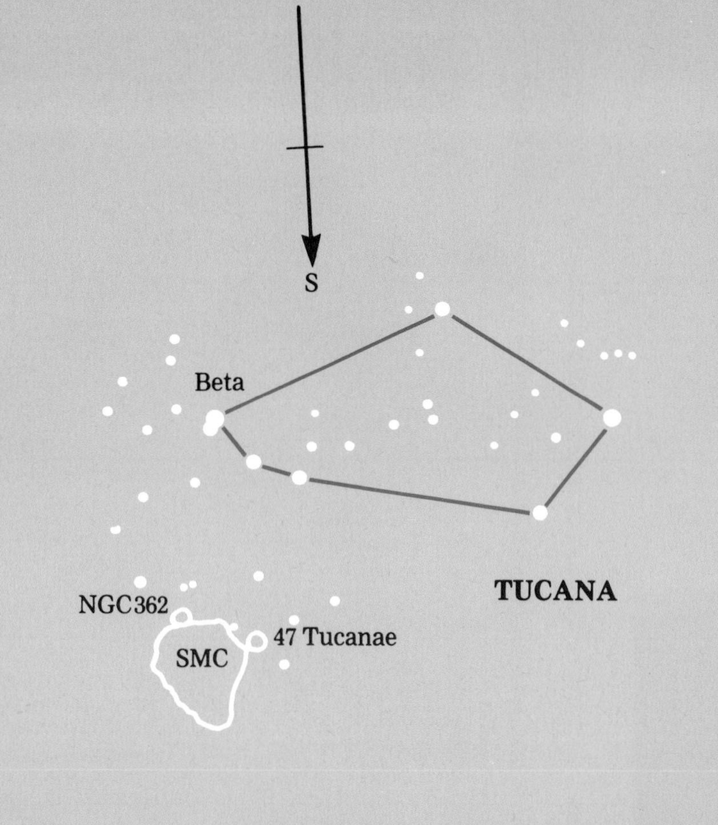

***Left:** Tucana contains a number of interesting objects, including the two globular clusters 47 Tucanae and NGC 362. However, most notable is the Small Magellanic Cloud, an irregular galaxy situated at a distance of around 200,000 light years. This system is plainly visible to the naked eye and some individual stars can be resolved through binoculars. Within the Small Magellanic Cloud many different types of object have been detected, including variable stars, star clusters and nebulae. In the picture (**below**) four bright concentrations of stars can be seen towards the upper left. These are (from the left) NGC 395; NGC 371, a large cluster with a diameter of around 300 light years; NGC 346, a huge nebulous mass of stars almost 600 light years long, and NGC 330, a cluster with a diameter of over 100 light years. Like its nearby companion the Large Magellanic Cloud in Dorado, the S.M.C. is a member of the Local Group of galaxies.*

__Right:__ this photograph shows the starfield around Beta Tucanae. Beta 1 and Beta 2 form a nice double with magnitudes of 4.52 and 4.48 separated by 27.1 seconds of arc, PA 170°. Beta 3, at magnitude 5.16, rounds off a pretty trio of stars visible in a small telescope. Each of these three stars is double, although the companions can only be resolved through large telescopes. The Beta system is seen against an attractive backdrop of stars.

__Left:__ this photograph covers a wider field of view than that above and takes in the area around the Small Magellanic Cloud and the two nearby globular clusters 47 Tucanae and NGC 362. These clusters are not physically related to the S.M.C. but they lie in the same line of sight. 47 Tucanae shines at 5th magnitude and is clearly visible to the naked eye. Telescopes of 4 inches aperture or more will resolve individual stars. 47 Tucanae lies at a distance of around 16,000 light years and has a diameter of more than 200 light years. Slightly fainter at 6th magnitude is NGC 362, which can be glimpsed with the naked eye under dark, clear skies. Both globulars make attractive binocular objects.

The Southern Sky: Summer

During summer the Milky Way is well placed and many magnificent star fields await the attention of the backyard astronomer. In the north it passes through Perseus, low down on the horizon, Auriga and between Orion the Hunter and Gemini the Heavenly Twins. After Monoceros it drifts through Puppis and Vela, Crux and Centaurus, marking its path down towards the southern horizon. Canopus is high in the sky, rivalled by Sirius the Dog Star in Canis Major. Sirius is found by extending upwards the line formed by the three stars in Orion's Belt, although its brilliance renders it unmistakable.

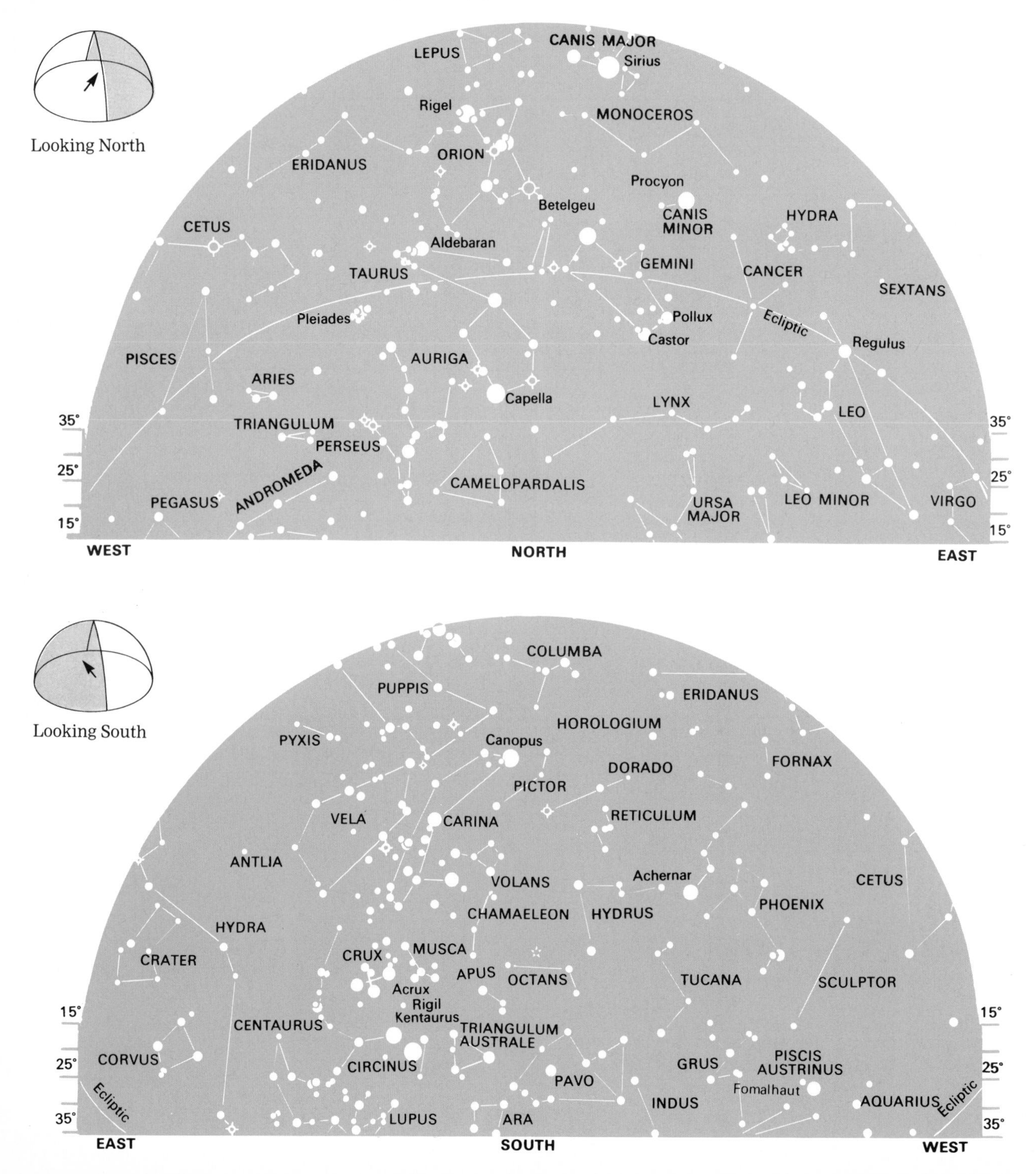

Below: *the globular cluster NGC 1851 in Columba. At 7th magnitude it can be seen in binoculars as a circular patch of hazy light.*
Right: *Mu Columbae is seen to be travelling through space with a velocity equal to 0.025 seconds of arc per year. Although its motion is difficult to detect from Earth-based observations except over long periods, it is equivalent to an actual speed of around 75 miles per second. The star seems to have been ejected from a region of nebulosity in Orion a few million years ago, although the reasons for this are still somewhat obscure.*

ORION

LEPUS

Mu Columbae

COLUMBA

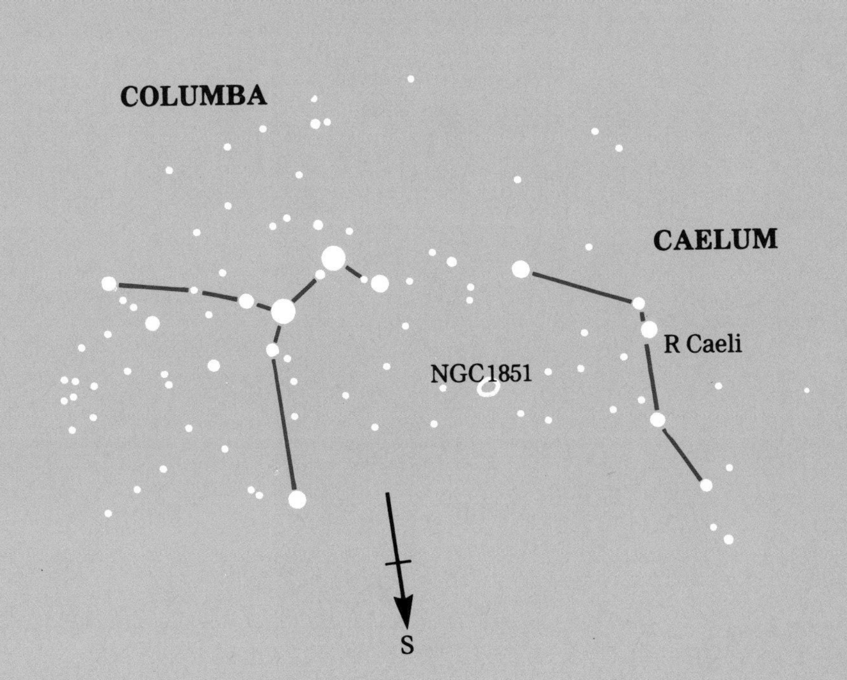

Left: *Columba and Caelum lie in a fairly barren region of sky to the southwest of the constellation Canis Major. An object of interest in Caelum is the long-period variable star R Caeli, which varies between 7th and 14th magnitude over a period of around 391 days. At minimum it is well beyond the reach of either binoculars or small telescopes, although it is interesting to watch it slowly fade from view, only to reappear a few months later as it climbs back to maximum brightness.*

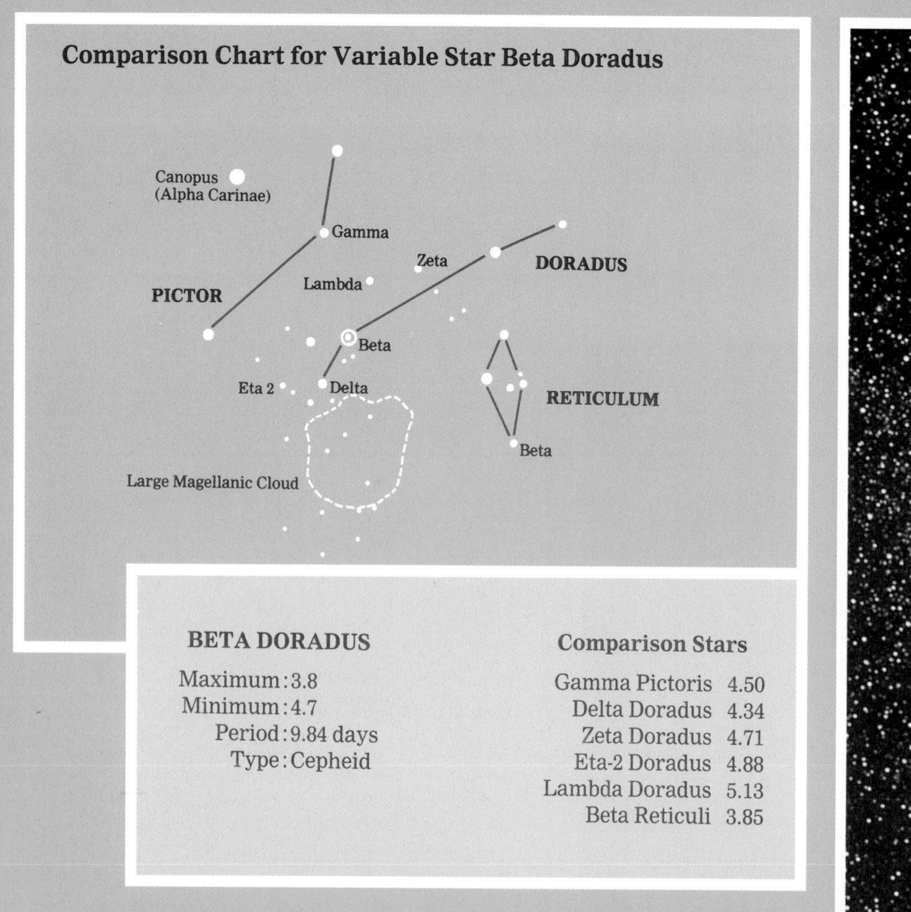

BETA DORADUS

Maximum: 3.8
Minimum: 4.7
Period: 9.84 days
Type: Cepheid

Comparison Stars

Star	Magnitude
Gamma Pictoris	4.50
Delta Doradus	4.34
Zeta Doradus	4.71
Eta-2 Doradus	4.88
Lambda Doradus	5.13
Beta Reticuli	3.85

Above left: *Beta Doradus is a Cepheid-type variable star which lies at a distance of around 1,700 light years. When at maximum brightness its light output is somewhere in the region of 7,000 times that of our Sun. The chart shows suitable comparison stars, while the photograph* (***left***) *shows the general field of Pictor, Doradus and Reticulum with Beta Doradus visible just above centre. The photograph also shows the Large Magellanic Cloud, an irregular galaxy situated at a distance of around 190,000 light years, making it marginally closer than the nearby Small Magellanic Cloud in Tucana.*

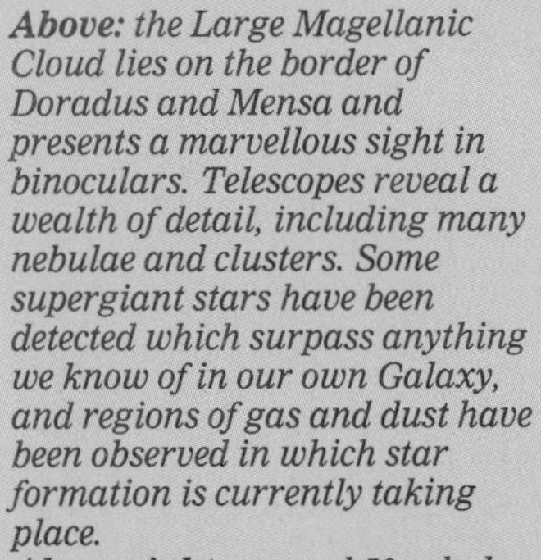

Above: *the Large Magellanic Cloud lies on the border of Doradus and Mensa and presents a marvellous sight in binoculars. Telescopes reveal a wealth of detail, including many nebulae and clusters. Some supergiant stars have been detected which surpass anything we know of in our own Galaxy, and regions of gas and dust have been observed in which star formation is currently taking place.*
Above right: *around 50 nebulae have been observed in the Large Magellanic Cloud, the brightest of which by far is the Tarantula Nebula (NGC 2070). This is a gigantic diffuse nebula some 800 light years in diameter which is actually visible to the naked eye, despite its distance of some 190,000 light years. This colossal mass of glowing gas is, as far as we know, unequalled anywhere in the Universe.*

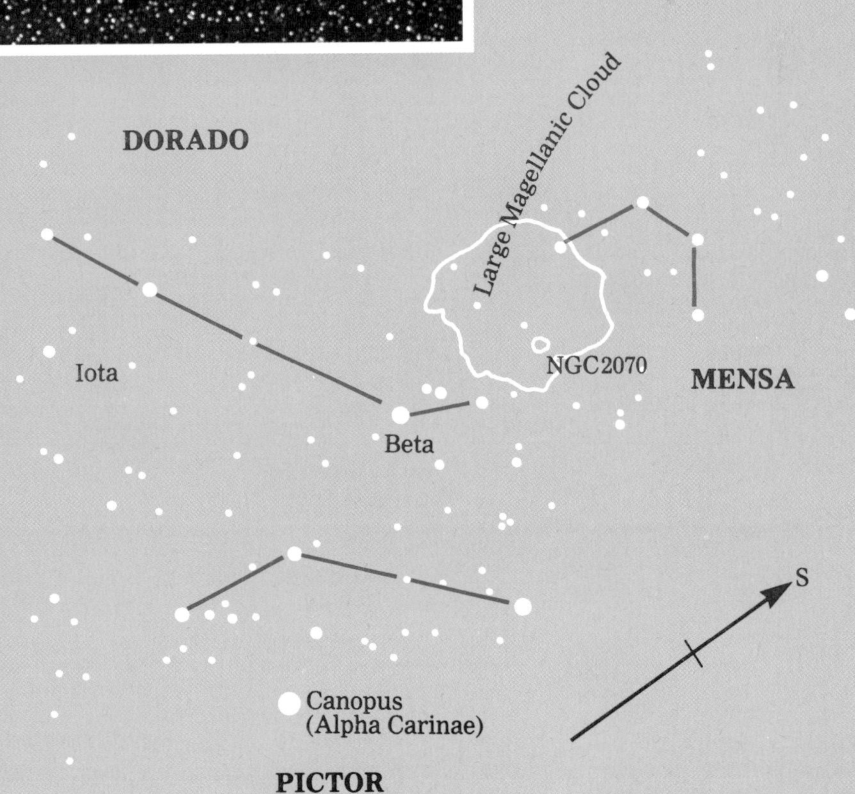

The Southern Sky: Autumn

The northern aspect is dominated by the Zodiacal groups of Gemini, Cancer, Leo and Virgo, although the bright stars of Orion are still making their presence felt low over the horizon. For southern hemisphere observers, this is the best time to see Ursa Major, the Great Bear, which can be found prowling low over the northern horizon. The curve of stars in the Bear's tail leads the way to Arcturus in Boötes and Spica in Virgo. Looking towards the south the Milky Way is prominent. Look also for Crux and Centaurus and, in particular, for Omega Centauri, the brilliant naked-eye globular cluster.

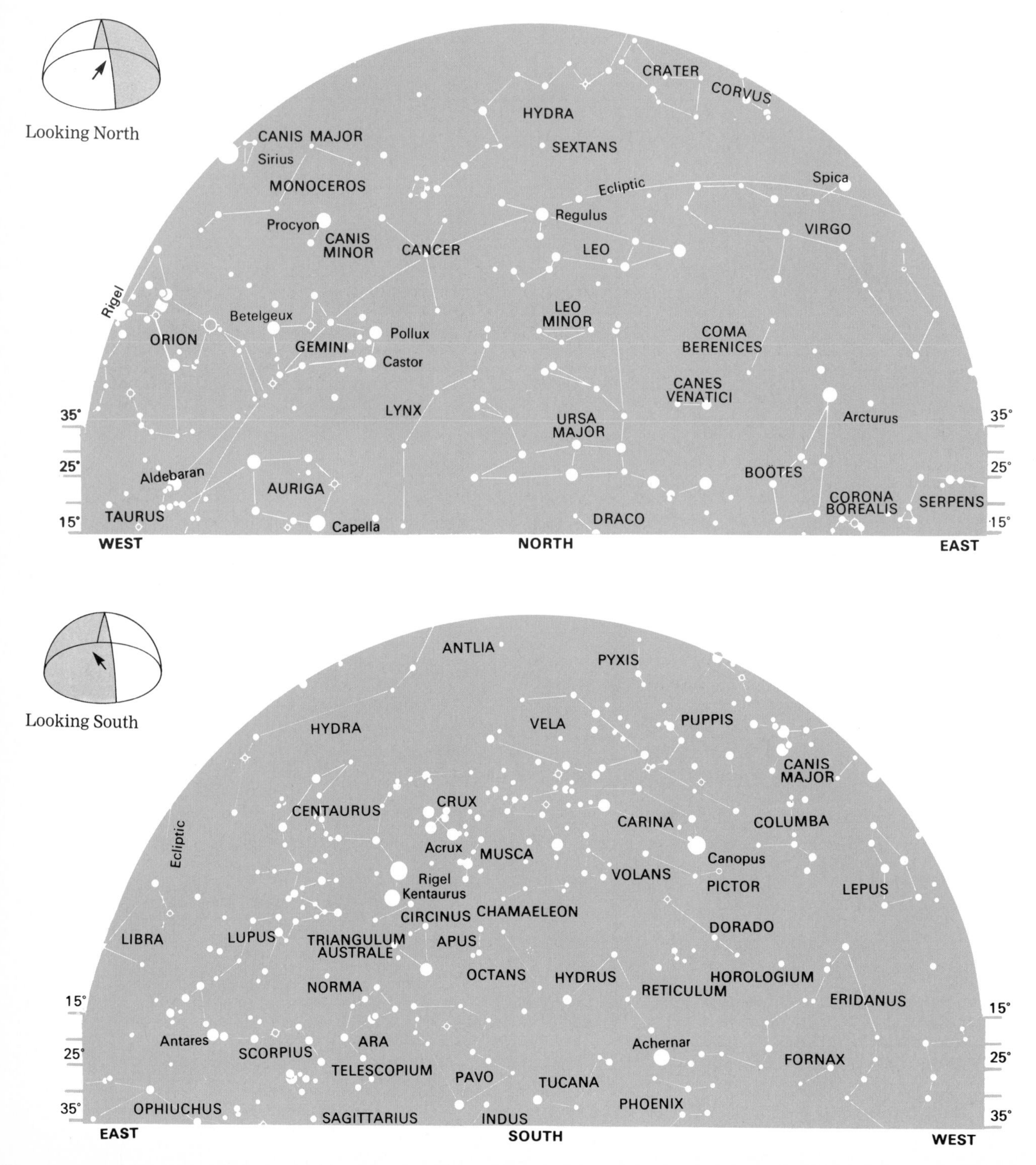

Left: *the photograph shows the area of Musca and Crux with part of the adjoining constellation of Centaurus. Musca, although a small group, has quite a distinctive shape and is easily picked out. The pair of stars Mu and Lambda Muscae offer quite a nice colour contrast in binoculars with the white of Lambda bringing out the distinctly red tint of Mu. Look also for the slightly orange hue of Epsilon.*

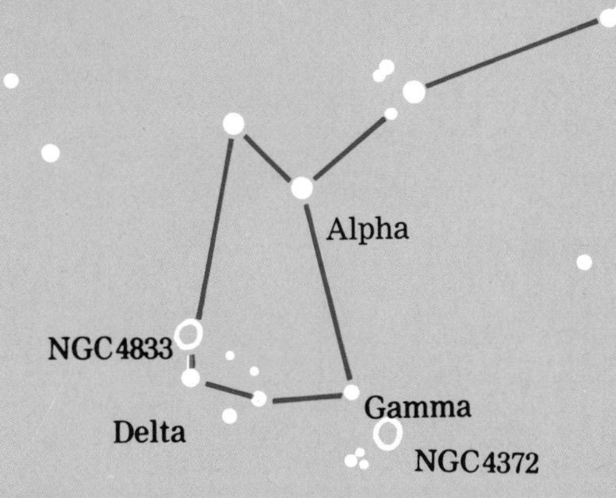

Right: *Musca contains a number of star clusters, including the two globulars NGC 4372 and NGC 4833. Both are of 8th magnitude and can be seen in binoculars as tiny, circular discs of light. The open cluster NGC 4463, found on a line between Alpha Muscae and Alpha Crucis, is quite faint but can be picked up in binoculars or a small telescope. Apertures of at least 6 inches will be needed in order to resolve stars in any of the above clusters.*

Left: this photograph shows most of Centaurus, including the two bright stars Alpha Centauri and Agena (Beta Centauri). Alpha is interesting because it is the closest of the bright stars, shining with a magnitude of −0.27 from a distance of 4.3 light years. The dim red dwarf Proxima, just to the southwest of Alpha, is thought to be slightly closer, although at magnitude 10.7 telescopes will be required to see it. Also seen in this picture is the globular cluster Omega Centauri.

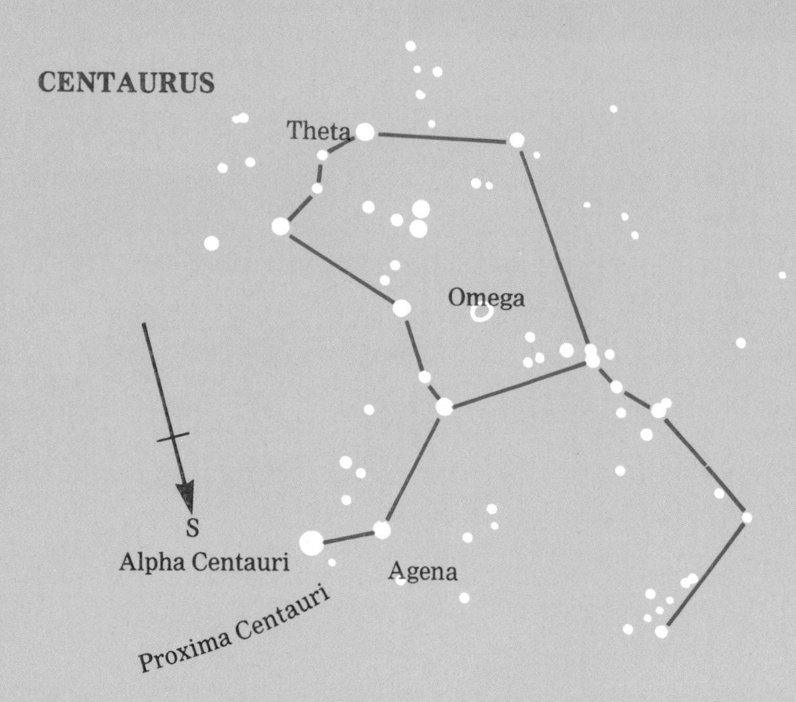

Above: wide-field photograph of the Centaurus region.
Left: chart of Centaurus showing the location of Omega Centauri, the finest globular cluster in the heavens. Shining at 4th magnitude it is clearly visible to the naked eye and was even recorded by the Greek astronomer Ptolemy almost 2,000 years ago. Omega Centauri lies at a distance of around 17,000 light years and is thought to contain up to a million stars. Some outlying stars can be seen through binoculars, although wide-field telescopes give stunning views of this magnificent cluster. The most northerly bright star in Centaurus is Theta, which may be seen to have an orange tint through binoculars. The Milky Way passes through the southern regions of Centaurus and the area is well worth sweeping with either binoculars or a wide-field telescope.

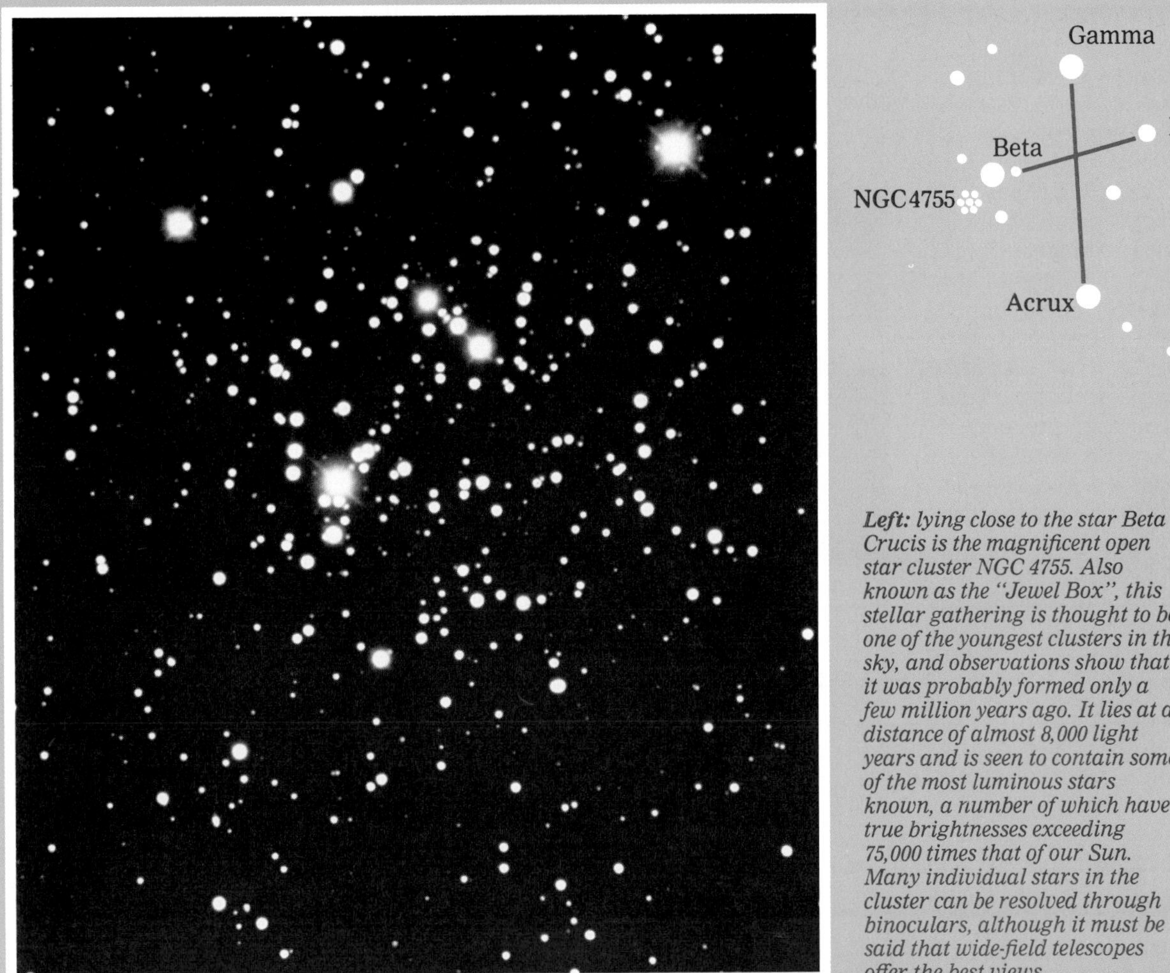

Left: *lying close to the star Beta Crucis is the magnificent open star cluster NGC 4755. Also known as the "Jewel Box", this stellar gathering is thought to be one of the youngest clusters in the sky, and observations show that it was probably formed only a few million years ago. It lies at a distance of almost 8,000 light years and is seen to contain some of the most luminous stars known, a number of which have true brightnesses exceeding 75,000 times that of our Sun. Many individual stars in the cluster can be resolved through binoculars, although it must be said that wide-field telescopes offer the best views.*

Right: *this photograph shows Crux, "Southern Cross", rising over the Indian Ocean. The brightest star in the constellation is Acrux, which shines with a magnitude of 0.87, making it the 14th brightest star in the sky. Acrux is actually a binary with a very long orbital period. The two components are of magnitudes 1.39 and 1.86 and currently lie 4.5 seconds of arc apart. The northernmost bright star in Crux is Gamma, which can be seen to have a slight orange-red tint. This is in contrast to the other main members of the group, which are all white. Gamma is a giant star, the light from which has taken around 220 years to reach us. Its true luminosity is 900 times that of the Sun.*

Professional Observatories

Until quite recently the visible light emitted by the stars and other celestial objects has been the only source of information available to us. Now that has all changed. Visible light is just one form of electromagnetic radiation and today's astronomer must study the radiation emitted throughout the entire electromagnetic spectrum in order to build up a clearer and more complete picture of the Universe. There are many types of radiation that are unable to penetrate the Earth's atmosphere, and to carry out research at these wavelengths it is necessary for the astronomer to take his instruments out into space. Observation carried out at wavelengths outside the visible portion of the electromagnetic spectrum is commonly referred to as invisible astronomy, and its growth in recent years has been rapid.

Gamma-ray, X-ray and ultraviolet radiations are all absorbed by the atmosphere, and in order to carry out research at these wavelengths satellites are used. Before the advent of the satellite, observation of these types of radiation was made by a variety of means. Helium-filled balloons carried detectors to heights of between 28 miles (45 km) and 31 miles (50 km), where they were able to detect ultraviolet emissions, together with a certain amount of gamma radiation. High-altitude sounding rockets, many of which were launched in the late 1940s and early 1950s, attained heights of 62 miles (100 km) or so and carried out measurements of X-ray emissions. The greater the distance above the Earth's surface, the more "transparent" the atmosphere becomes, and the next logical step for astronomers, once rocketry had developed enough to enable satellites to be placed into orbit, was to send detectors permanently above the atmosphere. The big advantage of this was that observation time was not restricted to short periods as with previous methods, and comparatively large amounts of data could be collected.

Examination of infra-red and radio emissions is made more easy by the fact that radiation at these wavelengths can pass through our atmosphere. Whereas X-rays and gamma-rays are absorbed by individual gas

Left: *the Anglo-Australian Observatory at Siding Spring, New South Wales.*
Right: *the Infra Red Astronomy Satellite (IRAS) made this image of the Andromeda Galaxy at a number of different wavelengths. Blue represents the warmest material, green is material at intermediate temperatures and red the coldest. A bright ring of yellow cloud shows regions where star formation is taking place.*
Below: *the 250-ft (76 m) dish at Jodrell Bank, Cheshire, is one of the world's largest fully-steerable radio telescopes.*

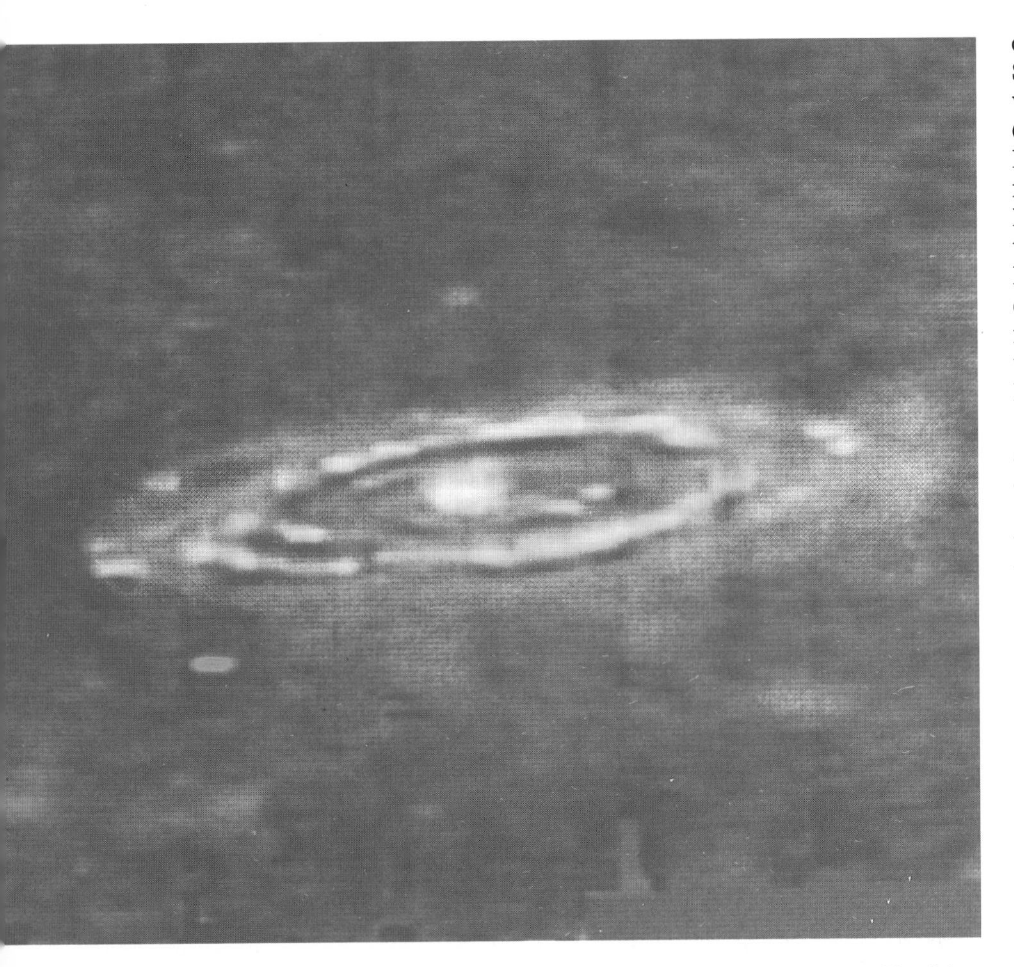

atoms and molecules within the atmosphere and ultraviolet by molecules in the ozone layer, infra-red radiation falls prey to water vapour. It follows, therefore, that infra-red telescopes will function if situated above the water vapour in the atmosphere. An example is the 92-inch (2.34 m) diameter Wyoming Infrared Telescope, located at a height of around 9,850 feet (3,000 m) on Jelm Mountain near Laramie. However, infra-red telescopes have also been put into orbit, a notable example being the highly successful Infra Red Astronomy Satellite (IRAS), which carried out a survey of the sky at infra-red wavelengths during 1983.

Radio astronomy can be carried out at sea level as virtually all the radio emissions from the sky can penetrate our atmosphere. It is only the longest wavelengths that are unable to reach ground level, being reflected back into space from the ionosphere. Radio telescopes are found all over the world, and one of the largest fully steerable instruments is the 250-feet (76 m) Mark 1A dish at Jodrell Bank, Cheshire. However, not all radio telescopes are fully steerable, and into this category comes the huge 1,000-feet (305 m) Arecibo radio telescope, built into a natural hollow in hills near Arecibo, Puerto Rico. Radio astronomy was born in 1931, when the American engineer Karl Jansky first detected radio waves from the sky. It has come a long way since then, and radio astronomers are at the forefront of our attempts to probe the deepest reaches of the universe.

Optical Astronomy

In spite of the growth and importance of invisible astronomy, the optical astronomer still has a major role to play, a fact emphasized by the emergence of observatory complexes at Mauna Kea, Hawaii, and the La Palma Observatory in the Canary Islands. Mauna Kea is home to a large array of telescopes, including a 144-inch (3.65 m) reflector operated as a joint effort between Canadian and French astronomers and the University of Hawaii. Situated on top of an extinct shield volcano at a height of 14,000 feet (4,300 m) above sea level, the Mauna Kea site is one of the best in the world for the observational astronomer. However, it is the opinion of many that the world's best site is that of the El Roque de los Muchachos Observatory on La Palma. British astronomy is really coming to the fore on La Palma with the refurbished 100-inch (2.5 m) Isaac Newton Telescope and the giant 165-inch (4.2 m) William Herschel Telescope helping British astronomers explore the skies.

The largest optical telescope in the world is the 236-inch (6 m) diameter reflector at Zelenchukskaya Observatory in the Soviet Union, although for many years the honour belonged to the 200-inch (5 m) Hale Reflector at Palomar Observatory in America. Completed in 1948, the Hale Reflector was the brainchild of George Ellery Hale, who had already been responsible for the construction of a number of other fine instruments, including the 100-inch (2.5 m) "Hooker" Telescope, completed in 1917 and situated at the nearby Mount Wilson Observatory.

The southern hemisphere also has its fair share of astronomical observatories, particularly in Australia, where southern skies come under the watchful eye of telescopes at the Siding Spring Observatory near the little town of Coonabarabran. The main telescope at Siding Spring is the 153-inch (3.9 m) Anglo-Australian Telescope. Inaugurated in 1974, this 335-ton optical masterpiece has produced some marvellous results, including what are probably the best ever photographs of the night sky.

The attentions of the professional astronomer can be focused in many directions. However, whether research is carried out with radio telescopes, from orbiting detectors or through optical instruments, the information gleaned from each region of the electromagnetic spectrum will be put together and analysed to produce a clearer understanding of the universe in which we live.

Amateur Observatories

Small telescopes – say refractors of up to 75 mm (3 in) aperture, or reflectors of up to 100 mm (4 in) – are usually quite portable, easily stored indoors and taken outside when needed. Even relatively bulky refractors of 100 mm (4 in) and reflectors of 150 mm (6 in) can be moved about, albeit awkwardly. But instruments larger than this are far too portly to carry without the risk of either straining yourself or damaging the telescope. A permanent, secure and weatherproof outdoor observatory is a must.

Types of observatory
There are basically three formats of observatory you could build: the run-off shed; the run-off roof; and the revolving dome. Each is really little more than a robust garden shed in construction, with adaptations to permit the use of the telescope.

Domed observatory
By far the best type – but also the most difficult to construct – is the archetypal domed observatory. It offers full protection to both observer and telescope in addition to minimizing the detrimental effects of stray lighting. The rotating dome need not be solid – a timber frame clad with roofing felt would do – but it must be leak-proof, and able to withstand high winds.

A slide-away observatory
The walls of a run-off shed may split in half and both portions parted to reveal the telescope, or the structure might feature a panel that, once removed, allows the complete construction to be pulled away from the instrument.

In either case the movable sections must be fitted with wheels that run on metal rails set into the base of the observatory. In its disfavour it offers no weather protection at all to user or telescope when in use.

A removable roof observatory
The run-off roof, ideal for tall instruments such as refractors or large reflectors, is probably the easiest amateur observatory to build (and the type of construction illustrated overleaf). A pent-roofed building, the sloping roof panel simple slides off to reveal the sky to the viewer. Its four walls help to shield the telescope and observer from the wind and most stray lighting.

A suitable site
Vital to any observatory is a flat, solid base on which to mount the telescope – a cast concrete slab on firm foundations is essential. Perching the instrument on top of a garage or other building would subject the instrument to vibrations, and expose it to wind and other elements. Also, warm air currents rising from below would severely distort the telescopic image with an effect similar to air distortion seen just above a tarmac road during hot weather.

An isolated site is preferable: aim to build the observatory as far away from other buildings as possible. An unobstructed horizon would also be ideal, although unlikely in a built-up area, where houses and street lights would distract. Even trees can interrupt observation.

Above: *one of the buildings at the Norman Lockyer Observatory is the McClean Dome, which houses both a 10-inch visual and a 12-inch photographic refractor. The Norman Lockyer Observatory was built in 1913 and was originally used by professional astronomers, although it is now available for amateur use. The Sidmouth Astronomical Society carries out much work there.*
Left: *Rosemary Naylor's run-off shed houses a 10-inch reflector. The telescope is exposed by removing the end wall and wheeling the rest of the building to one side.*

Above: Derek Aspinall's observatory at Sowerby Bridge, near Halifax, is rectangular and provides useful shelf space in the corners of the rotating upper section.
Left: West Yorkshire Astronomical Society's observatory at Pontefract has a conventional dome beneath which are a meeting room and an adjoining dark room.

For general observing in the northern hemisphere a clear southern aspect is preferable, as all the celestial objects visible from your latitude will at some time pass through the region of sky between the north celestial pole and the southern horizon. For observers in the southern hemisphere the opposite is the case; an unobstructed northern horizon is preferred as all celestial objects will pass between the south celestial pole and the northern horizon.

Calculating the dimensions
When you have chosen a suitable site for the observatory, consider its vital statistics. At least 300 mm (12 in) of wall clearance should be allowed on all sides: assume that the telescope will be located in the centre of the floor and measure the distance from the centre of the mount to the end of the tube opposite the mirror. Add 300 mm (12 in) to this to give the minimum distance from the centre of the observatory to the inside of one wall. Double this to arrive at the internal dimensions of the building.

To determine the roof height, set the telescope to its "rest" position (with the tube horizontal) and measure the distance from the floor to the highest point on the telescope, taking into account the finder and eyepiece mount. This height should correspond to the height of the lowest wall of the observatory.

In theory, the lower the walls are the better – it isn't essential to allow for personal headroom, as the roof is removed during use – but you may want to cater for a larger telescope in the future.

For maximum sky exposure for observers in the northern hemisphere, site the lowest wall at the southern end of the building (vice versa for those in the southern hemisphere).

To calculate the height of the highest wall of the observatory, consider the angle of inclination of the roof. For an observatory measuring, say, 2.75 × 2.75 m (9 × 9 ft) with the lowest wall 1,500 mm (60 in) high, a height difference of 300 mm (12 in) would be suitable – giving an inclination of 1 in 9 – to enable the roof to be slid off easily.

Making your own Observatory

This pent-roofed garden observatory comprises timber wall frames – which you can clad cheaply with second-hand floorboards or man-made boards – with a slide-off roof glazed with corrugated plastic sheeting that's lightweight, durable and weatherproof. The whole structure is mounted on a cast concrete base, vital to avoid vibrations. The following instructions refer to a typical structure with internal dimensions of 2.75 × 2.75 m (9 × 9 ft).

What you need to build the Observatory

Tools and equipment

strings and pegs
spade
sledgehammer or roller
hammer
screwdriver
spirit level

Materials

- concrete (1 part cement; 2½ parts sand; 4 parts coarse aggregate).
- hardcore.
- four 150 × 25mm (6 × 1in) formwork planks.
- four 100mm sq (4in sq) sawn softwood corner posts.
- eight rag bolts, washers and nuts.
- sixteen lengths 100 × 50mm (4 × 2in) softwood for wall frames (plus extra length for door lining).
- two 25mm sq (1in sq) softwood wheel guides.
- T&G cladding for walls and door (or marine plywood sheets).
- eight lengths 75 × 50mm (3 × 2in) softwood for roof frame.
- sheets of corrugated PVC roofing.
- 38mm (1½in) galvanized nails.
- 50mm (2in) galvanized nails.
- 75mm (3in) galvanized nails.
- 50mm (2in) No. 8 countersunk screws.
- four nylon wheels.
- four handles.
- two door hinges.
- one door handle and lock.
- four padlocks and hasps.

Casting the base slab

1 *Mark out the site with strings stretched between wooden pegs. Clear and level the ground leaving a base of firm well-compacted subsoil. On soft ground, add a well-rammed-down layer of hardcore.*

2 *At each corner of the base, dig or bore a 450mm (18in) deep hole roughly 300mm (12in) sq to take posts of 100mm sq (4in sq) preservative-treated sawn softwood. The posts should be equal to the height of the observatory walls (in this case 1.5m/5ft or 1.8m/6ft) plus the distance from the bottom of the holes to the level of the base. Ideally the tops of the posts should be bevelled to match the angle of the roof slope.*

3 *Secure the posts vertically in the holes with temporary props, standing the posts on bricks for drainage, and pour concrete around them (buy this pre-mixed in bags). Check that the posts are set squarely by measuring the diagonals, which should be identical. Tamp the mix to expel air then cover with polythene and leave to cure.*

4 *Nail lengths of 150 × 25mm (6 × 1in) softwood across pairs of posts to make the form for the concrete floor. Make sure they are set level and that the tops of the form boards are flush with the intended finished surface of the slab.*

5 *Add approx. 50mm (2in) of hardcore and ram down, then spread on a blinding layer of sand or fine gravel to fill any holes.*

6 *You'll need about 0.75 cu m (24 cu ft) of concrete for a slab about 100mm (4in) thick in the size described here. It's simplest to buy ready-mixed concrete, which is delivered and dumped on site ready to be spread and compacted level to the top of the formwork. Use a stout plank drawn aross the surface with a chopping motion then drawn back again in a sawing motion to achieve the level.*

7 *While the concrete is wet push pairs of rag bolts into the surface between posts as anchor points for the wall frames. Check that the rag bolts are set evenly apart and are vertical and that 75mm (3in) of thread protrudes above the surface. Now leave the slab to harden completely for a few days.*

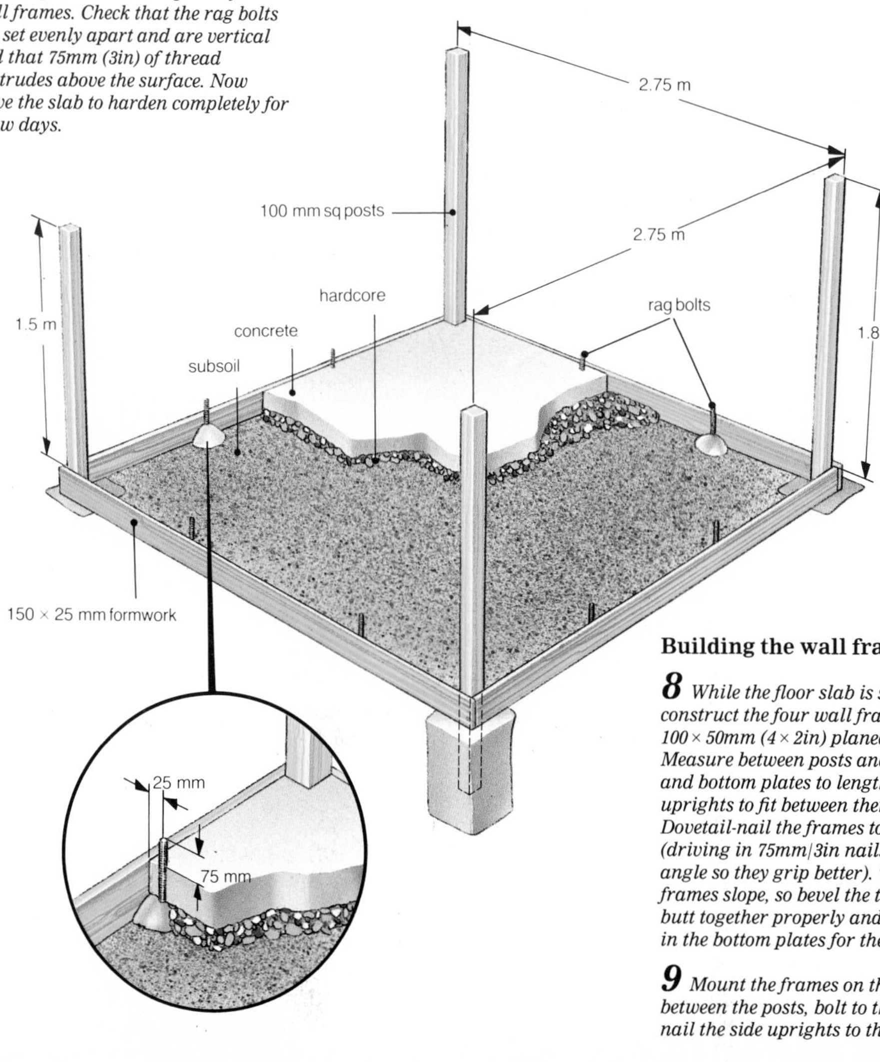

Building the wall frames

8 *While the floor slab is setting, construct the four wall frames, using 100 × 50mm (4 × 2in) planed softwood. Measure between posts and cut the top and bottom plates to length; cut the uprights to fit between them. Dovetail-nail the frames together (driving in 75mm/3in nails at an angle so they grip better). The side frames slope, so bevel the tops so they butt together properly and drill holes in the bottom plates for the rag bolts.*

9 *Mount the frames on the slab between the posts, bolt to the base and nail the side uprights to the posts.*

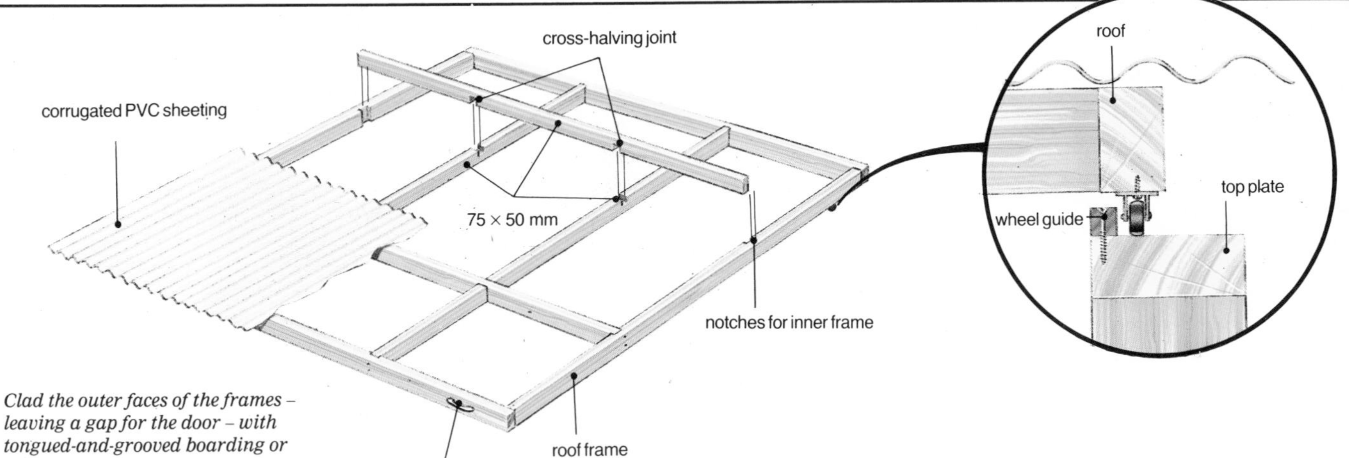

Clad the outer faces of the frames – leaving a gap for the door – with tongued-and-grooved boarding or panels of marine plywood, and nail to the top and bottom panels using 50mm (2in) nails.

10 *Skew-nail an extra upright to the top and bottom plates as a door lining. Make a door from strips of T&G boarding braced with horizontal and diagonal battens then hinge to the door lining.*

Fitting the roof

11 *The roof comprises an outer framework of 75 × 50mm (3 × 2in) softwood, the components butt-jointed and nailed. An inner frame comprising four pieces connected by cross-halving joints is fitted into notches cut in the inner face of the perimeter frame. Mark out and saw the joints with all lengths clamped together. Use an offcut to mark the joint width lines; the joints are half the depth of the wood.*

12 *Assemble the outer frame then assemble and fit the inner frame, tapping home with a mallet. Reinforce each joint with 38mm (1½in) nails.*

13 *Use panels of transparent corrugated PVC to clad the roof frame, screwing them to the struts and outer frame using the special screws and washers provided; fit the screw caps over the heads. Work from the lower edge of the roof to the top, overlapping sheets by about 150mm (6in). Overlap panels in width by two corrugations for a watertight seal. Profiled foam strips are available for sealing the ends. The panelling should overlap the outer sides by two corrugations and the top and bottom by about 100mm (4in).*

14 *Screw pairs of handles to the top and bottom of the roof frame. The roof runs on small nylon wheels, two screwed to the top edge of the underside of the roof panel itself and two screwed to the top of the side walls. Nail a pair of wheel stops about 50mm (2in) from the wheels on the side walls. Make runner guides by screwing two strips of 25mm sq (1in sq) softwood along the inner top edge of the side walls.*

15 *When slid back the lower edge of the roof can rest on a post pivoted onto the ground. Next, bolt a 50 × 25mm (2 × 1in) timber batten to the back of one of the downside posts to swivel as a temporary roof stop.*

16 *Four padlocks and hasps fasten the roof to the walls, positioned about 300mm (12 in) from each end of the side walls.*

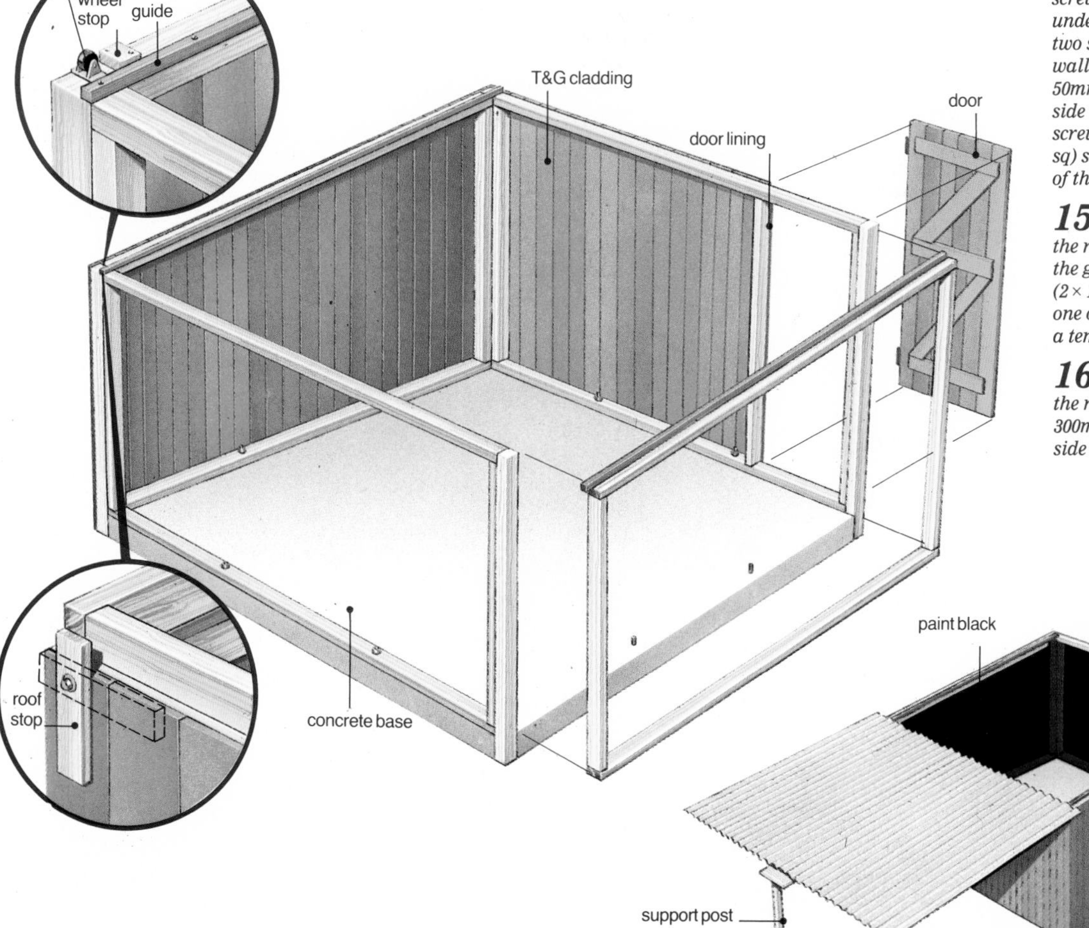

Recording Observations

Whenever observations are made it is desirable to make a record of what is seen. For absolute beginners this can take the form of a straightforward list of the stars and constellations identified. For the more advanced observer they could detail such things as the differing positions of Jupiter's moons, the changing aspect of Saturn's rings or the march of sunspots across the solar disc. As well as being interesting to look back on, these recordings may provide a useful source of information for observing projects. It doesn't take long to get into a routine of noting what you see, although observations should be recorded as soon as possible after they are made while they are fresh in the memory. Many can be noted at the time, exceptions being drawings and sketches, which can be completed shortly afterwards.

One of the best methods of recording observations is in an observing book. This doesn't have to be anything elaborate or expensive – a simple exercise book is ideal. It is useful to have a clipboard to hold the book when in use to ensure that it remains flat and steady. Some form of illumination can be attached to the clipboard, powered by batteries strapped to the back. This will help immensely in that it will keep one hand free to make adjustments to the telescope and so on. It is important, however, that a red light should be used as this will have the least effect on your night vision. Also, if prolonged observing sessions are planned, don't forget to wrap up well. If in doubt, put on more clothes than you think you'll need.

For each observation there are a number of details that should be noted as follows:

Date and Time: The "double date" should always be used. For example, if an observation is made on the evening of 25 February, 1987, the entry should read: 1987, February 25/26. Following this should be a note of the exact time of the observation, given in terms of Greenwich Mean Time (GMT), otherwise known as Universal Time (UT).

Observation: Jupiter
Date: 14/15 Sept 1986
Time (GMT): 00:20
Observing Site: Shipley, West Yorkshire, England.
Instrument: 222mm f6 reflector
Magnification: 178 X
Conditions: Partial cloud. Large clear breaks.
Limiting Magnitude: Mag 5·6
Seeing: Antoniadi III

Observation: Saturn
Date: 29/30 May 1986
Time (GMT): 21:23 GMT
Observing Site: Shipley, West Yorkshire, England
Instrument: 222mm f6 reflector
Magnification: 267 X
Conditions: clear but slightly hazy
Limiting Magnitude: mag 5
Seeing: Antoniadi II

There are instances when corrections may be necessary to the time shown on your watch. For example, when observing in Britain during the summer, it is important to take into account British Summer Time (BST), which differs from Greenwich Mean Time by one hour. Also, observations made in other parts of the world should take the local time system into account. Generally speaking, the astronomical publications available in each country outline the method for converting to Greenwich Mean Time. GMT operates on a 24-hour system running from midnight to midnight, e.g. 21h 36m GMT.

Left: *a back garden plays host to a group of dedicated amateur astronomers as they gaze at the wonders of the night sky. You needn't travel far to enjoy the sights that our Universe has to offer, be it through naked-eye star-spotting, sweeping the sky with binoculars or probing deep space with telescopes.*

Instrument used: If the observation was made with the naked eye then simply record the fact. If binoculars are used then give their specification (7 × 50, 10 × 50 and so on). With telescopes the aperture and type of telescope, together with the magnification used, should be noted (80 mm refractor @ 120 ×, 150 mm reflector @ 72 × and so on).

Observing site: If possible, note the latitude and longitude of the observing site. It is useful to find out these details for the most commonly used site, be it an observatory, back garden or otherwise. Failing that, at least give the geographical location.

Observing conditions: State whether there is any moonlight or if observing is done under twilight conditions. The presence of any cloud or mist should also be noted, as should the "limiting magnitude", which is the magnitude of the faintest star visible to the naked eye. This gives a good indication as to the clarity of the atmosphere. Finally, note any adverse effects that air turbulence may have on the quality of image produced by the telescope. The famous astronomer Eugene Michael Antoniadi devised a scale that described various grades of "seeing" conditions and the effects they had on image quality. This scale is now widely used and is particularly useful when different groups of astronomers are comparing results, or when an individual observer wishes to compare his own observations with some made on a previous occasion. Conditions according to the Antoniadi Scale should be indicated by a Roman numeral to signify one of the following grades:

I Perfect seeing, without a quiver.
II Slight undulations, with moments of calm lasting several seconds.
III Moderate seeing, with large air tremors.
IV Poor seeing, with constant troublesome undulations.
V Very bad seeing, scarcely allowing the making of a rough sketch.

Whatever the type of observation, sketch, draw or record only what you actually see, and not what you want to see. If you are unsure about the visibility of any particular feature or otherwise, by all means make an additional note in the observations to that effect. Any records kept will be far more valuable if they are made as accurately as possible and are a true account of your efforts.

Astrophotography

To make a start in astrophotography you need a camera, cable release, tripod and suitable film. The camera should be a single lens reflex (SLR) and capable of long exposures. Most SLRs have a "B" shutter setting allowing indefinite exposures. A cable release allows the shutter to be operated without shaking the camera; and a tripod holds the camera steady. For astrophotography, a fast film is required for good results in dim conditions. Film speeds are normally indicated by either an ASA or DIN number – the higher the number, the faster the film. One of the best for beginners is 400 ASA (27 DIN).

Films can be obtained that produce either prints or slides, each of which can be in colour or black and white. Generally, black and white films give better image contrast. In the case of a black and white print film, the negatives make ideal slides for projecting on to a screen and can show a great deal of detail. Whatever film you use, when you take it in for developing state clearly that it contains pictures of the night sky. Also, if you have slide film developed, ask for the transparencies to be supplied in an uncut roll to avoid some careless individual cutting your prized astrophotos in half just because they didn't see where the picture margins were! It is a simple job to mount your own slides.

The basic equipment now assembled, remove the lens cap and make sure that the camera focus is set on infinity (∞) and that the aperture is as wide open as possible. (The smaller the number, the wider the aperture setting and the more light reaches the film.) The only decision now is what to photograph, and there are a wealth of alternatives.

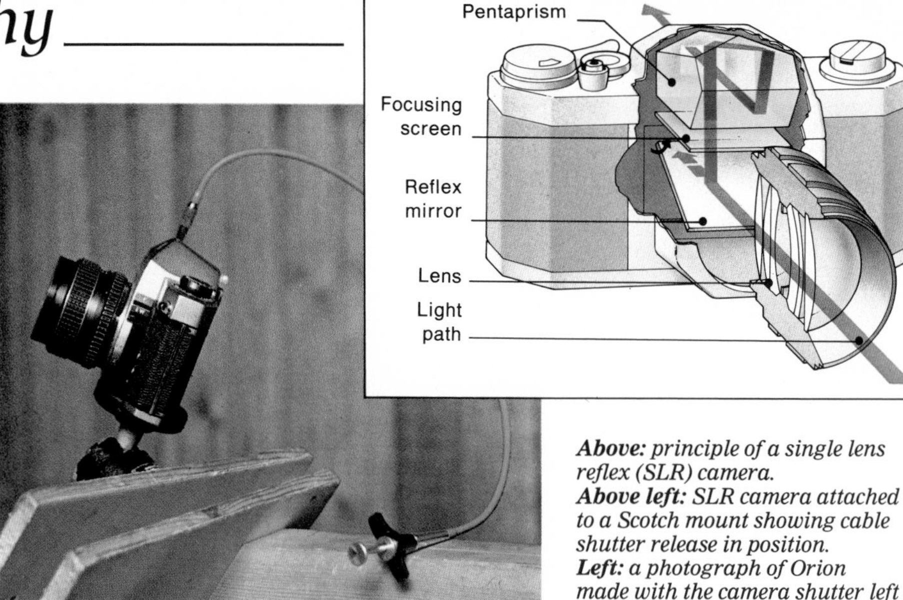

Above: *principle of a single lens reflex (SLR) camera.*
Above left: *SLR camera attached to a Scotch mount showing cable shutter release in position.*
Left: *a photograph of Orion made with the camera shutter left open throughout. An initial short exposure captured star images, after which the lens was carefully covered for a short time with a dark cloth. Care must be taken not to shake the camera at this stage. The cloth is then removed, a longer exposure then being made to produce the star trails. The camera shutter is then closed.*

Star trails

Because the Earth turns on its axis the stars and other celestial objects appear to rise in the east, travel across the sky, and set in the west. This movement from east to west is a result of the Earth spinning from west to east. In a photograph taken over several minutes the stars will appear not as points but as streaks of light, commonly known as star trails. A clear dark sky is essential and you must be as far away as possible from any artificial lighting, which, together with moonlight, will fog the film very quickly, making long exposures useless.

Point the camera at a region of the sky containing a number of bright stars or a well-known constellation, open the shutter using the cable release, and take exposures of 5 minutes', 10 minutes' and 15 minutes' duration. The darker the sky, the longer the exposure that can be made.

Now point the camera at the Pole Star, or Polaris, and carry out a similar sequence. Because the Earth's axis points towards Polaris, the stars appear to travel around it, and photographs taken this way will show circumpolar star trails.

Constellations

To obtain photographs of constellations that show the stars as points of light, exposures should be much shorter. Again, pick out a bright group of stars and make a number of exposures. These should be 10 seconds, 20 seconds and 30 seconds long. Anything longer than 30 seconds will produce star trails.

The Moon and Planets

A good time to take shots of the Moon and planets is when they appear in the twilight sky (see bottom picture on page 22). At these times the Moon will be seen as a crescent, and when one or more of the planets, such as Mercury or Venus, happen to be in the same region, photographs can be taken showing these celestial gatherings. In these situations the sky will not be completely dark, and exposures should be no longer than 30 seconds, and often less.

Meteors

At certain times of the year meteor showers occur during which the number of observed meteors can rise dramatically. Each shower appears to radiate from a particular point in the sky, and your best chance of photographing a meteor is to direct your camera at a position roughly 20 to 30 degrees away from this radiant. Leave the shutter open for as long as you can without the film becoming fogged by background light that may be present. Experience will tell you how long this will be for various conditions. Close it and repeat the procedure for as long as necessary, or until a meteor crosses the field of view. If this happens, close the shutter immediately in order to keep the meteor trail bright and clear on the photograph. It is by no means easy to photograph a "shooting star". Although you may capture one straightaway, patience may be required before you succeed.

Artificial Satellites

On any clear night the occasional satellite may be seen as a point of light crossing the sky. They move quite slowly and can take up to two or three minutes to travel right across the heavens. Point the camera at a region of the sky through which the satellite will pass, and hold the shutter open until the satellite passes from view.

All the above techniques can be carried out using a standard lens, so don't try to run before you can walk. Experiment with these ideas before you attempt to move on to more sophisticated equipment and more complicated projects.

Left: *photograph taken on 4 September 1983, showing an artificial satellite passing through Ursa Major.*
Below: *a group of observers (complete with warm clothing and sun-loungers) carrying out a meteor watch. Cameras are also mounted for meteor photography.*

Making a Simple Astrophotography Mount

Of the astrophotographic projects you could experiment with, few are as enthralling as capturing deep-sky objects – star clusters, nebulae and galaxies, for instance – on film.

However, you will be hampered using only a stationary camera: apparent movement of celestial objects – actually due to the rotation of the Earth – causes star images to appear as bright streaks on exposures longer than about 30 seconds.

Photographing galaxies such as the Milky Way may demand exposures as long as 10 to 15 minutes (overall sky conditions and the level of artificial light permitting), so if the stars are to stay in the same place on the film, the camera must be moved continuously during exposure in a direction counter to the Earth's rotation. An equatorial, or Scotch, mount, is a device which rotates on an axis pointed at the celestial pole – parallel to the Earth's axis. Professional motor-driven mounts are available for lenses greater than 1,200 mm, but you can improvise a hand-driven mount for use with shorter lenses.

What you need to make an Equatorial Mount

Materials

- one planed softwood anchor block, 400 × 100 × 50 mm
- two plywood panels, 320 × 140 × 12 mm
- one planed softwood camera arm, 150 × 35 × 25 mm
- two steel plates, 50 × 20 × 1 mm
- one steel pointer gauge, 60 × 20 × 1 mm
- one right-angle bracket, 20mm wide × 3 mm thick, with arms 100 mm and 75 mm long
- one brass heavy-duty butt hinge, 100 mm long
- three M6 bolts (two 62 mm long, one 50 mm long) plus two brass washers, three plain nuts, and two wing nuts to fit
- four M4.5 countersunk bolts (two 19 mm long, two 30 mm long), plus nuts to fit
- ten 12 mm long No. 6 woodscrews
- two 30 mm long No. 10 countersunk woodscrews
- one strong elastic band

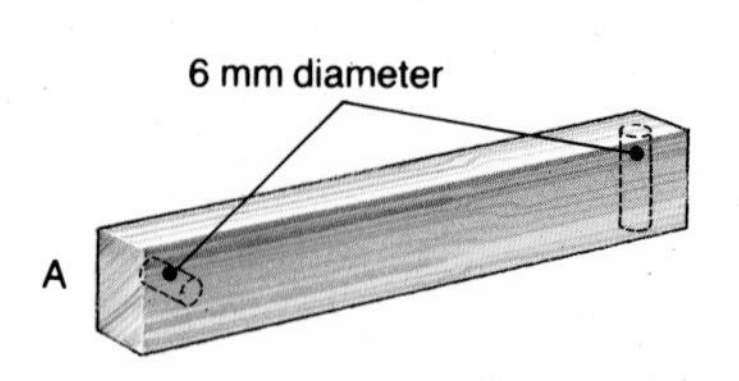

Drill two 6mm bolt holes through the camera arm, at right-angles to each other, about 19 mm from each end (A).

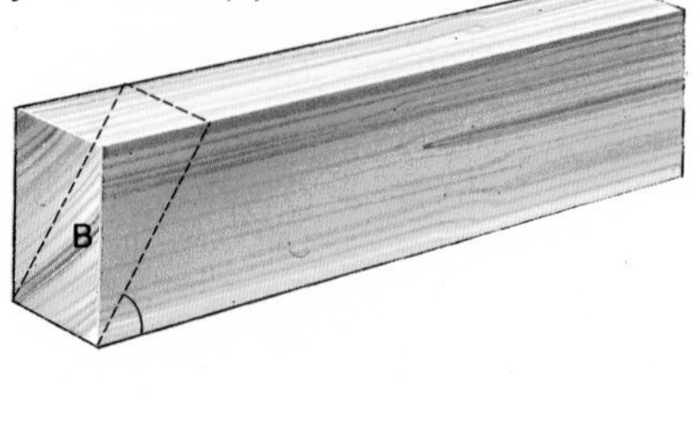

Tools

Try square; protractor; rule; saw; screwdriver; power drill with bits suitable for wood and metal; countersinking bit; metal-working file; chisel; adjustable spanner

Use a protractor to mark the angle equal to the latitude of the location on a length of 100 × 50 mm timber, 400 mm from a squared end. Cut the piece to length to form the anchor block (B).

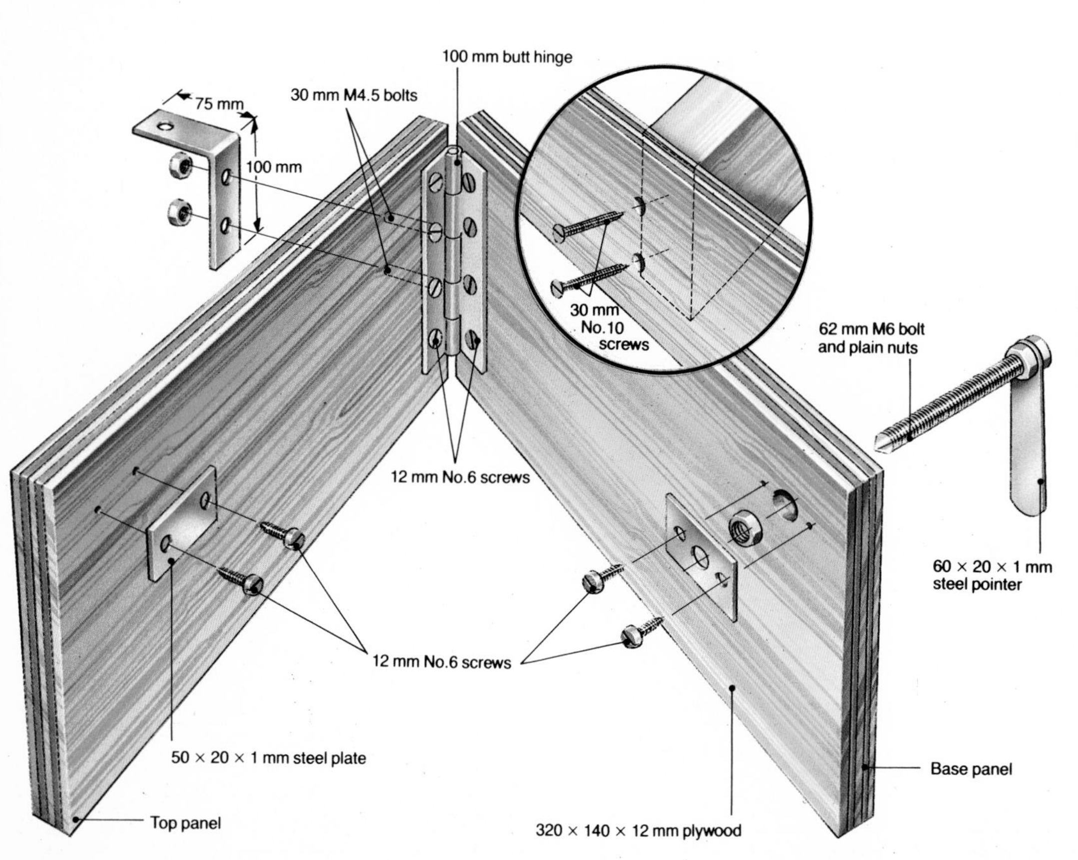

How the mount is made

The mount illustrated comprises a sturdy wooden anchor block, with one end cut at an angle that corresponds to the latitude of the location. A base panel is screwed across the angled end and a second panel hinged to one end of the first.

The camera is fixed via a machine bolt to a wooden arm, which pivots on – and can be locked to – a right-angled bracket screwed to the outer face of the hinged panel, near the hinge end. Angular movement of the hinged panel is by a manually operated 1 mm-thread drive screw with a pointer gauge attached, which pushes against a metal plate fixed to the underside of the panel. A strong elastic band wrapped around the boards keeps the hinged board at the correct tension.

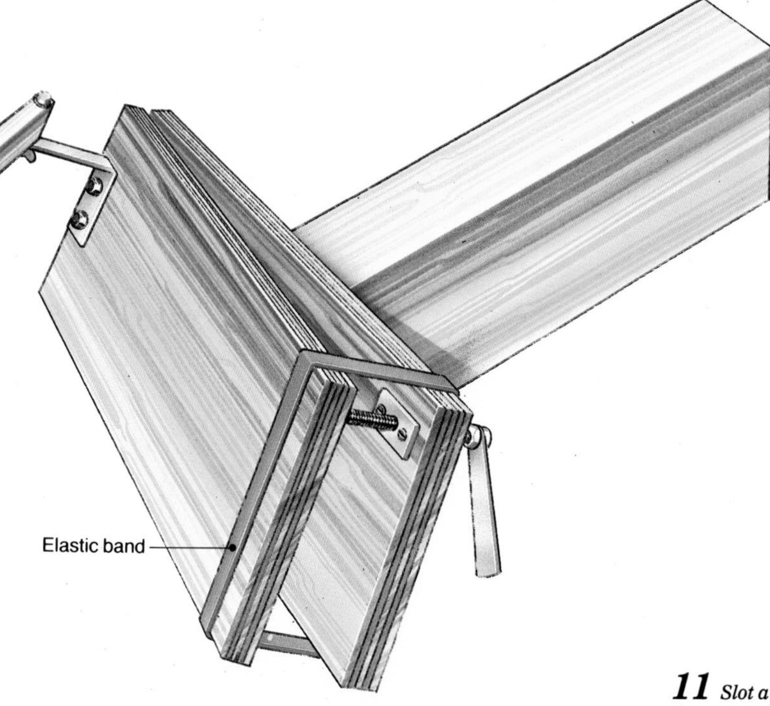

Assembling the Mount

__1__ Mark out and cut top and base panels then, on one, mark, drill and countersink two 5 mm clearance holes for No. 10 woodscrews, 40 mm in from each side of the panel, and halfway along, for attaching the panel to the anchor block.

__2__ Lay the hinge on the base panel, its spine running flush with the end, and mark the fixing hole positions. Remove the hinge and drill 1.5 mm pilot holes then return the hinge and secure to the base panel with No. 6 screws.

__3__ Measure and mark a point on the base panel 40 mm in from one long edge and 229 mm along from the central axis of the hinge. Drill a 6 mm hole through the panel and chisel a recess to take a plain nut; tap the nut in flush with the face of the bottom panel.

__4__ Scribe a cross at the centre of each metal plate and drill two 4 mm holes in two of the 50 × 20 mm metal plates, about 6 mm from each end. Drill a 6 mm hole in the centre of one plate. Hold this plate in place on the base panel, mark then drill 1.5 mm pilot holes and screw over the nut lodged in its recess using No. 6 screws. Mark the position of the second metal plate on the underside of the top panel so it corresponds with that of the fixed plate on the base panel. Drill pilot holes then screw into place.

__5__ Lay the panels end-to-end, aligning the top panel flush with the spine of the hinge. Mark through the fixing holes then drill clearance holes at the two central positions to take the two M4.5 bolts, which also secure the bracket. Drill 1.5 mm pilot holes for screws at the two remaining outer fixing positions.

__6__ If you cannot buy a bracket of the correct size, make one by bending a strip of metal to form 75 mm and 100 mm long arms. Drill two 4.5 mm holes in the long arm corresponding to the two bolt holes drilled in the top panel, and a 6 mm diameter hole near the end of the short arm.

__7__ Secure the top panel and bracket to the hinge using the M4.5 bolts, slotted through from the hinge side, and tighten their nuts with a spanner. Fit two No. 6 screws at the remaining fixings.

__8__ To make the drive screw assembly, first drill a 6 mm diameter hole near one end of the 60 × 20 mm metal strip – which can have its sharp corners rounded off – and pass a 62 mm long M6 bolt through. Secure the strip with a plain nut. Screw the bolt into the nut housed in the base panel leaving about 12 mm protruding. File the end of the bolt to an even point, rounding off the tip slightly.

__9__ Place the base panel on the angled end of the anchor block and align their lower edges. Mark the anchor block through the pre-drilled holes in the panel and drill 2 mm pilot holes. Attach the panel to the block with two No. 10 screws.

__10__ Slot a brass washer onto a 62 mm long M6 bolt then push the bolt through one of the holes in the camera arm and through the hole in the end of the metal angle bracket fixed to the mount assembly. Slip another washer onto the bolt and secure with a winged nut. Swivel the camera arm around and lock at any position by tightening the wing nut.

__11__ Slot a 50 mm long M6 bolt through the hole in the other end of the camera arm, secure with a plain nut and screw on the remaining wing nut, inverted. There should be no more than 4.5 mm of bolt protruding from the base of the wing nut – sufficient to screw into the threaded bush in the base of the camera. Lock the camera onto the mount by tightening the wing nut.

__12__ Loop the strong elastic band around the hinged and base boards and turn the drive screw carefully to open and close the hinged board.

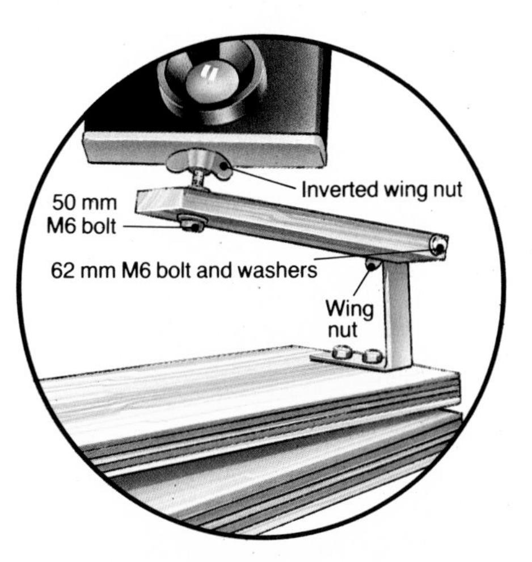

Using the mount

- Align the anchor block horizontally on a north-south line, with the main assembly at the southern end, clamped (say, to a portable workbench) or weighted down so that it is rigid; the hinge axis points towards the north celestial pole.
- Focus the camera on the section of sky you want to photograph by turning it on the arm, locking it, then pivoting the arm on its bracket and locking that when the picture is composed.
- Set the relevant aperture and film speed, then depress the shutter release. Immediately turn the drive screw at a rate of one revolution per minute: check that the guide pointer keeps pace with the second hand on a watch.

Glossary

Absolute magnitude The *apparent magnitude* that a *star* would have if observed from a distance of 10 *parsecs.*
Alt-azimuth mounting A type of telescope mounting that allows the instrument to be moved in two axes; horizontally (parallel to the horizon) and vertically (at right angles to the horizon).
Apastron The point in its orbit around a star that a body is at its furthest from the star.
Aperture The clear diameter of a telescope *objective.*
Aphelion The point in its *orbit* at which a body is furthest from the *Sun.*
Apogee The point in its orbit around the Earth that a body is at its furthest from the Earth.
Apparent magnitude The brightness of a *star* or other telestial object as seen from Earth.
Asteroid See *minor planet.*
Autumnal Equinox The point at which the *ecliptic* crosses the *celestial Equator* as the *Sun* passes from north to south.

Binary Stars Two *stars* in *orbit* around each other and held together by their mutual gravitational attraction.
Black hole The region of space surrounding an object that is so dense that even light cannot escape its gravitational field. These objects are thought to represent the final evolutionary stage of very massive *stars.*
Bolometric magnitude The magnitude of a *star*, taking into account all radiation emitted by it.

Celestial equator The projection of the Earth's equator onto the *celestial sphere.*
Celestial poles Imaginary points on the *celestial sphere* directly in line with the Earth's axis.
Celestial sphere The imaginary sphere upon which the *stars* and other celestial objects are found.
Circumpolar stars *Stars* which are so close to the *celestial poles* that they never set. The number of circumpolar stars varies according to the latitude of the observer.
Comet An object composed of a mixture of gas, dust and ice which travels around the *Sun* in an *orbit* that is generally very eccentric.
Conjunction The position whereby two celestial objects are lined up with each other as seen from Earth. A *planet* is at superior conjunction when situated at the other side of the *Sun* as seen from Earth and at inferior conjunction when located directly between the *Sun* and Earth.
Constellation One of a total of 88 patterns of *stars* found on the *celestial sphere.*

Declination The angular distance between a celestial object and the *celestial Equator* (abbr. dec.).
Double star Two *stars* which lie in almost the same line of sight as seen from Earth and which therefore seem to be close to each other.

Eclipse The total obscuration of one celestial body by another. An eclipse of the *Sun* occurs when the *Moon* passes directly between it and the Earth. An eclipse of the *Moon* takes place when the Earth passes between the *Sun* and *Moon*, and the Earth's shadow is thrown onto the lunar surface.
Ecliptic The apparent path of the *Sun* through the sky during the course of a year. The band of constellations through which it passes is known as the *Zodiac.*
Ellipse A closed, oval-shaped form, obtained by cutting through a cone at an angle inclined to the main axis of the cone. The *planets* all have elliptical *orbits* around the *Sun.*
Equatorial mount A type of telescope mounting that allows the instrument to follow the path of a *star* or other celestial object by adjustment around one axis only.
Eyepiece A small system of lenses attached to a telescope through which the eye looks. The magnification of the telescope can be altered by changing the eyepiece.

Galaxy A vast collection of *stars*, which also contains large amounts of gas and dust, and which measures many *light years* across. There are some galaxies, known as elliptical galaxies, which are composed mainly of *stars* with very little interstellar gas and dust.
Gas planet One of the four largest *planets* – Jupiter, Saturn, Uranus and Neptune – which are made up mainly of material in a gaseous or liquid form around a small, possibly rocky core.

Inferior planet A *planet* that travels around the *Sun* in an *orbit* inside that of the Earth.

Light year The distance, approximately equal to 6,000,000,000,000 miles (10,000,000,000,000 km), that is travelled by light in a year, and which is a standard measurement used by astronomers.
Limb The edge of the visible disc of an object such as the *Sun* or *Moon.*

Meridian An imaginary circle that crosses the *celestial sphere*, passing through both *celestial poles.*
Meteor The streak of luminescence seen in the sky which results from a *meteoroid* entering the Earth's atmosphere and burning up through friction with air particles.
Meteorite A *meteoroid* which is of sufficient size to at least partially survive the passage through the Earth's atmosphere.
Meteoroid The general term applied to interplanetary meteoritic debris.
Milky Way The faint band of light crossing the sky which is made up of the combined glow from thousands of stars which lie along the plane of our Galaxy. The Milky Way Galaxy is the name given to the gigantic spiral-shaped *galaxy* of which the *Sun* is a member.
Minor planet One of a large number of relatively tiny planetary bodies which orbit the *Sun* generally between the *orbits* of Mars and Jupiter. The largest, Ceres, is only just over 625 miles (1000 km) in diameter.
The Moon The Earth's only natural *satellite.*

Nebula An interstellar cloud of gas and dust. There are three types of nebula: reflection nebulae shine by reflecting the light of nearby stars; emission nebulae shine by absorbing ultraviolet radiation from nearby stars and re-emitting it as visible light; and dark nebulae are those which contain no stars and which appear as dark clouds, blotting out the light from the stars beyond.
Neutron star An object which results from the collapse of a *star* of between 1.4 and 3 *solar masses*.
Nova A *star* which is a member of a *binary system* and which undergoes a sudden increase in brightness.

Objective The light-gathering component of a telescope which takes the form of a lens, in the case of a refracting telescope, or a mirror in the case of a reflecting telescope.
Occultation The passage of one celestial body in front of another as seen from Earth, such as occurs when the *Moon* passes in front of a *star*. Strictly speaking, an *eclipse* is a type of occultation.
Opposition The point at which a *superior planet* is directly opposite to the *Sun* in the sky and which is therefore due south at midnight.
Orbit The path of one object around another.

Parallax The angular shift of a nearby star in relation to the background of more distant stars when viewed from two positions.
Parsec The distance, equal to 3·26 *light years*, at which a *star* would undergo an angular shift of one second of arc.
Penumbra The area of partial shadow which appears during a solar *eclipse* and which is seen to surround the *umbra*, or darker central region of shadow. It is also the name given to the lighter part of a *sunspot* surrounding the dark *umbra*.
Periastron The point in its orbit around a star that a body is at its closest to the star.
Perigee The point in its orbit around the Earth that a body is at its closest to the Earth.
Perihelion The point in its *orbit* at which a body is nearest to the *Sun*.
Perturbation The change in orbital path of a body by the gravitational influence of another body.
Planet One of the nine major members of the Sun's family.
Prime Meridian The *meridian* that cuts through the *celestial Equator* at the *Vernal Equinox*.
Pulsar A rapidly spinning *neutron star* which emits regular bursts of radiation.

Quasar An extremely distant and highly luminous object, thought to represent the nuclei of active *galaxies*.

Right ascension Measurement from west to east along the *celestial Equator*, expressed in hours, minutes and seconds. This measurement is taken from the point at which the *ecliptic* crosses the *celestial Equator* as the *Sun* moves from south to north. This point is known as the *Vernal Equinox*.

Satellite A small object in *orbit* around a much larger parent body.
Sidereal period The time taken for an object to complete one *orbit* around another. For example, the sidereal period of Mars is 687 Earth days, which is the time it takes to travel once around the *Sun*.
Solar mass The mass of the *Sun*. The masses of certain celestial objects are sometimes expressed in solar masses. For example, the mass of the nearby Andromeda Spiral Galaxy is around 300,000 million solar masses.
Solar system The collective term applied to the *Sun*, together with the *planets, comets, asteroids*, planetary *satellites* and interplanetary debris in *orbit* around it.
Star An object that shines due to the release of energy produced at its core by nuclear reactions. This energy escapes as light and heat. The *Sun* is a typical star.
Sunspot A relatively cool, dark area on the solar surface.
Sun The *star* which is the central dominating member of the *Solar System*. It is the *Sun* which gives out the light and heat so essential to life on our planet.
Superior planet A *planet* that travels around the *Sun* in an *orbit* outside that of the Earth.
Supernova A colossal explosion that represents the destruction of a very massive *star* and which results in sudden and tremendous brightening.
Synodic period The interval between successive *oppositions* of an object in the *Solar System*. The term is also applied to the interval between similar phases of such an object.

Terminator The division between the light and dark hemispheres of a *planet* or *satellite*.
Terrestrial planet One of the four inner *planets* – Mercury, Venus, Earth and Mars – which is rocky in composition. It is thought that Pluto is similar in its make-up.
Transit The passage of one object across the face of another. For example, a transit of Mercury occurs when the planet crosses the solar disc as seen from Earth. The term is also applied to the passage of an object, such as a *star*, across a particular point, such as a *meridian*.

Umbra See *Penumbra*.

Variable star A *star* which varies in brightness. An effect which may be due either to changes within the star itself or to the passage of a faint member of a *binary* system in front of a brighter component. Variable stars of this nature are eclipsing binaries.
Vernal Equinox The point at which the *ecliptic* crosses the *celestial Equator* as the *Sun* passes from south to north.

White dwarf A small, dense *star* approaching the end of its life.

Zenith The point on the celestial sphere directly above the observer.
Zodiac The band of *constellations* through which the *ecliptic* passes.

Astronomical Reference

SELECTED BOOKS

General astronomy

The New Atlas of the Universe by Patrick Moore (Michell Beazley, UK and Crown, USA, 1984).
This large-format book represents a state-of-the-art account of our current knowledge of the Cosmos. It examines both the Solar System and the Universe beyond, and is well written and lavishly illustrated throughout.

Practical astronomy

A Field Guide to the Stars and Planets by Donald Menzel (Collins, 1980).
An extremely useful guide to the night sky which incorporates a photographic atlas of the heavens, constellation maps, and an atlas of the Moon. Applicable to observers in both northern and southern hemispheres.

The Skywatchers Handbook by Colin A. Ronan, Storm Dunlop and Brian Jones. Editor: Colin A. Ronan (Corgi Books/Marshall Editions, 1985).
A good overall guide for the practical observer. It outlines both the day and night sky and gives useful advice on how to observe what you see.

Burnham's Celestial Handbook by Robert Burnham, Jr (Dover Publications, 1978).
An invaluable in-depth guide to the Universe with marvellous details of each constellation and the celestial objects within them. In three volumes, this is one of the better observational aids on the market.

The Night Sky (Times Books).
An annual publication that gives the positions of the planets throughout the year. Details of forthcoming eclipses are also given. A useful and handy guide.

Astrophotography

Astrophotography for the Amateur by Michael Covington (Cambridge University Press, 1985).
A thorough introduction to astrophotography which tells how to record on film pictures of stars, galaxies, the Sun, Moon and planets, comets, meteors and eclipses. Written as a basic guide, it emphasizes the use of equipment that is readily available and easy to use.

Reference

Amateur Astronomer's Handbook by J.B. Sidgwick (Faber and Faber, 1971).
A comprehensive guide to the telescope and auxiliary equipment. This is a technical publication, although highly recommended for the serious amateur astronomer.

SELECTED SUPPLIERS

Broadhurst, Clarkson and Fuller,
Telescope House,
63 Farringdon Road,
London EC1M 3JB, England.
Suppliers of a wide range of telescopes and accessories, binoculars and observing aids.

Drake Educational Productions,
St Fagans Road,
Fairweather,
Cardiff CF5 3AE, Wales.
Astronomy wallcharts, filmstrips and slide/cassette sets on many aspects of astronomy and deep space.

Earth and Sky,
256 Bacup Road,
Todmorden,
Lancashire OL14 7HJ, England.
A large choice of astronomical books, slides, posters, colour and black and white prints, star atlases, planispheres, postcards, etc. Some NASA material also available.

Coulter Optical Company,
P.O. Box K,
Idyllwild,
California 92349, USA.
Medium to large aperture reflecting telescopes, good quality optics, telescope mounts.

TBR Optical,
P.O. Box 17129,
850 Hudson Avenue,
Rochester, New York 14617, USA.
Binoculars, telescopes, eyepieces, finders.

Astronomy,
Order Department,
1027 N. Seventh Street,
Milwaukee,
Wisconsin 53233, USA.
Books, charts, phanispheres, star atlases, observing guides. NASA material also available. *Astronomy, Deep Sky, Telescope Making* and *Odyssey* magazines.

Sky Publishing Corporation,
49 Bay State Road,
Cambridge,
Massachusetts 02238, USA.
Books, charts, star atlases and catalogues, observing guides. *Sky and Telescope* Magazine.

SELECTED ORGANIZATIONS

Federation of Astronomical Societies,
1 Tal-y-Bont Road,
Ely,
Cardiff, CF5 5EU, Wales.

Junior Astronomical Society,
22 Queensthorpe Road, Sydenham,
London, SE26 4PH, England.

British Astronomical Association,
Burlington House,
Piccadilly,
London, W1V 0NL, England.

American Association of Variable Star Observers,
187 Concord Avenue,
Cambridge,
Massachusetts 02138, USA.

British Astronomical Association
(New South Wales Branch),
Sydney Observatory,
Sydney,
NSW 2001, Australia.

Royal Astronomical Society of Canada,
136 Dupont Street,
Toronto,
Ontario, M5R 1V5, Canada.

Astronomical Society of Southern Africa,
South African Astronomical Observatory,
P.O. Box 9,
Observatory,
7935 Cape, South Africa.

Royal Astronomical Society of New Zealand,
P.O. Box 3181,
Wellington C1, New Zealand.

Credits

Photographs

Anglo Australian Telescope Board – 45R, 55, 70T
Ron Arbour – 78, 79B
Derek Aspinall – 99T
California Institute of Technology/Carnegie Institute – 44B, 45B, 47R, 71
Carl Zeiss, West Germany – 10L
Cerro Toloto Inter-American Observatory – 42B
CSIRO, Australia – 16B
J.R. Fletcher – 52B, 58TR, 65T, 68R, 69T
Peter Folford – 104B
The Hansen Planetarium, Salt Lake City – 70B, 77
Malcolm Johnson/Chris Walker – 50T, 53B, 73, 74
JPL/NASA – 24C
Link Observatory – 47BL
Peter McKanna – 99B
D.R. McLean – 27
A.H. Mikesell – 28
Paul Money – 24T, 30, 41, 50B, 51B, 54T, 57, 58TL&B, 59L, 63B, 65B, 66, 67, 69B, 75T, 79T, 105T
Patrick Moore – 96B
Mount Wilson Palomar Observatory – 46BR, 51T, 74T
NASA – 16T, 18, 22T, 26B, 44T, 47BC, 97
NASA/LRC – 21BL
Rosemary Naylor – 98B
NOAO – 16C, 43B, 86, 89, 90–91T, 95T
RAS – 46BL, BC, 46TL, TR, 47CL, c, 75B
Royal Observatory, Edinburgh – 26B, 34
Robin Scagell – 10R&B, 29TR&B, 36, 39, 49L&R, 52T, 53T, 54B, 61, 63T,68L, 76T, 81, 83, 85, 87, 90B, 93, 94, 95B, 96T, 98T, 102, 104T, 105B
Tasco Telescopes – 8, 9
Colin Taylor – 31
University of Arizona, Tucson – 20, 21T, C, BR
US Naval Observatory – 32, 43T, 45L, 47T, 59R, 60, 76B
Alan Young – 29TL, 38, 42T

Jacket Photographs – California Institute of Technology/Carnegie Institute, Tasco Telescopes

Illustrations

Julian Baum – 12/13
Andrew Booth – 103
Gerry Collins – 49–95
Peter Grego – 21
Trevor Lawrence – 8, 9, 38, 40R, 41
Stan North – 14BR, 17B, 22, 100–101, 106–107
Paul Williams – 11

General Index

Star Chart Index

For illustration credits see page 111